ZEROS OF POLYNOMIALS AND SOLVABLE NONLINEAR EVOLUTION EQUATIONS

Reporting a novel breakthrough in the identification and investigation of solvable and integrable systems of nonlinearly coupled evolution equations, this book includes many examples that illustrate this approach. Beginning with systems of ODEs, including second-order ODEs of Newtonian type, it then discusses systems of PDEs, and systems evolving in discrete time. It reports a novel, differential algorithm to evaluate all the zeros of a generic polynomial of arbitrary degree: a remarkable development of a fundamental mathematical problem with a long history. This book will be of interest to applied mathematicians and mathematical physicists working in the area of integrable and solvable nonlinear evolution equations; it can also be used as supplementary reading material for general Applied Mathematics or Mathematical Physics courses.

FRANCESCO CALOGERO is Professor of Theoretical Physics (Emeritus) at the Sapienza University of Rome, Italy. He has published numerous papers and books on physics and mathematics as well as on arms control and disarmament topics and he served as Secretary General of the Pugwash Conferences on Science and World Affairs from 1989–1997 (1995 Nobel Peace Prize).

ZEROS OF POLYNOMIALS AND SOLVABLE NONLINEAR EVOLUTION EQUATIONS

FRANCESCO CALOGERO

Physics Department, University of Rome "La Sapienza", Rome, Italy
INFN, Sezione di Roma 1

CAMBRIDGE
UNIVERSITY PRESS

CAMBRIDGE
UNIVERSITY PRESS

University Printing House, Cambridge CB2 8BS, United Kingdom

One Liberty Plaza, 20th Floor, New York, NY 10006, USA

477 Williamstown Road, Port Melbourne, VIC 3207, Australia

314–321, 3rd Floor, Plot 3, Splendor Forum, Jasola District Centre, New Delhi – 110025, India

79 Anson Road, #06–04/06, Singapore 079906

Cambridge University Press is part of the University of Cambridge.

It furthers the University's mission by disseminating knowledge in the pursuit of education, learning, and research at the highest international levels of excellence.

www.cambridge.org
Information on this title: www.cambridge.org/9781108428590
DOI: 10.1017/9781108553124

First published 2018

Printed and bound in Great Britain by Clays Ltd, Elcograf S.p.A.

A catalogue record for this publication is available from the British Library.

Library of Congress Cataloging-in-Publication Data
Names: Calogero, Francesco, 1935– author.
Title: Zeros of polynomials and solvable nonlinear evolution equations /
Francesco Calogero (Sapienza University of Rome, emeritus).
Description: Cambridge : Cambridge University Press, 2018. |
Includes bibliographical references.
Identifiers: LCCN 2018017272 | ISBN 9781108428590 (hardback)
Subjects: LCSH: Evolution equations, Nonlinear. | Zero (The number) |
Polynomials. | Differential equations, Partial. | Differential equations, Nonlinear.
Classification: LCC QC20.7.E88 C268 2018 | DDC 530.1/55353–dc23
LC record available at https://lccn.loc.gov/2018017272

ISBN 978-1-108-42859-0 Hardback

Contents

Preface

This short book is based on the *algebraic* relationship among the *N zeros* and the *N coefficients* of a (monic) polynomial of *arbitrary* degree *N*. This is a core mathematical topic since time immemorial; indeed the important mathematical subdiscipline called "algebraic geometry" originated from investigations of this relationship and is focused on its ramifications, which are quite deep. The approach used in this book is instead quite elementary; it is essentially based on the simple idea to consider (monic) polynomials of arbitrary degree *N* in their argument—say, the *complex* variable z—which moreover also depend on a parameter—say the *real* variable t—generally conveniently interpreted as "time" (the dependence on a second parameter playing the role of additional "space" variable is also considered, see Chapter 5). This implies that both the *N coefficients* and the *N zeros* of such a time-dependent monic polynomial *depend on time*, and this simple notion allows a certain number of interesting developments relevant to the topics mentioned in the title of this book. Some of these developments emerged quite recently; indeed this book is a compilation of results obtained and published by its author—alone or with collaborators—over the last three years. This is reflected in its bibliography and in the fact that occasionally the text below is drawn—even *verbatim*—from these publications. But unpublished findings are also included.

The author is very grateful to his co-authors—Oksana Bihun, Mario Bruschi, and François Leyvraz—who have been instrumental in obtaining some of the findings reported in this book (as specified below), and also to his co-authors of previous papers on related topics—several of which are identified in the bibliographic notes at the end of each chapter, and/or are referred to in the papers quoted there—including Fabio Briscese, Antonio Degasperis, Silvana De Lillo, Luca Di Cerbo, Riccardo Droghei, Marianna Euler, Norbert Euler, Jean Pierre Françoise, David Gómez-Ullate, Sandro Graffi, Vladimir I. Inozemtsev, Sabino Iona, Edwin Langmann, Mauro Mariani, Paolo Maria Santini, Matteo Sommacal, Ji Xiaoda, and Ge Yi. Interactions with these colleagues and with others I met in Rome and at international meetings over the years have been quite useful, in particular those facilitated by the series of international Gatherings of Scientists hosted—for three to five weeks in November-December every second year since 2000—by the Centro Internacional de Ciencias (CIC) in Cuernavaca, Mexico, as well as the series of Nonlinear Evolution Equations and Dynamical Systems meetings held since 1980 and the series of Physics and Mathematics of Nonlinear Phenomena meetings held since 1979 (the names of these meetings have evolved over time).

This book is written in plain language: it is addressed to a wide readership, including mathematicians, mathematical and theoretical physicists, and, more generally, practitioners interested in getting acquainted with mathematical tools useful to model complex phenomena via *systems* of *nonlinear* evolution equations—including *ordinary differential* equations, *partial differential* equations, and equations evolving via *discrete* time-steps. The phenomena for which the tools provided within this book might prove useful include many-body problems in classical mechanics, chemical reactions, ecology, population dynamics, economics, you name it. The readership of this book might moreover include students, indeed the results reported offer a fertile ground for master and PhD level dissertations exploiting the simple technique described herein to identify and investigate *new* systems of *nonlinear* evolution equations amenable to *exact* treatments. This book might be used—at the advanced undergraduate or at the graduate level—as basic text for a topical course or as background reading material for a pure and applied mathematics course where evolution equations play a significant role. It has been accordingly written so as to facilitate a *non-systematic* reading of it from beginning to end—at the cost of some repetitions (my teaching experience over more than half a century taught me that *repetita iuvant*). It is, however, advisable to read first Chapter 1 and especially Chapter 2; while the material in Chapter 3 should mainly appeal to *numerical* analysts, who might then hopefully become interested in investigating the comparative disadvantages/advantages—relative to traditional methods—of the *differential algorithm* to compute *all* the zeros of a *generic* polynomial reported there.

Finally, let us mention that throughout this book we do not hesitate from using *italics* whenever we consider this typographical device possibly useful to help the reader identify the *more significant* aspect of the text; our apologies to those readers who dislike this trick.

1

Introduction

As already mentioned above a main tool to identify and investigate *solvable* dynamical systems—including N-body models describing the evolution in the *complex* plane (or equivalently in the *real* Cartesian plane) of N point particles interacting *nonlinearly* among themselves and moving according to *Newtonian* equations of motion ("accelerations equal forces")—are the relations among the *time-evolutions* of the N *zeros* and those of the N *coefficients* of a monic *time-dependent* polynomial of (*arbitrary*) degree N. This approach was introduced decades ago [16] and has been extensively used since—see, for instance, the two books [20] and [27] and references therein—but the technique to implement it was essentially restricted to the consideration of the time evolution of the *zeros* of a (monic) polynomial the *coefficients* of which evolve according to a system of *linear* Ordinary Differential Equations (ODEs). The elimination of this restriction—to *linear* evolutions of the *coefficients*—was recently made possible by the key formulas reported in Chapter 2, opening thereby the way to the identification of many more *solvable/integrable* N-body problems, as described in Chapter 4 and demonstrated by the examples reported there. This approach has also been extended to the identification and investigation of *solvable nonlinear Partial Differential Equations* (PDEs: see Chapter 5) and to equations of evolution in *discrete time* (see Chapter 7).

Moreover—as described in Chapter 6—this development naturally led to the idea of *generations* of monic polynomials of degree N—with the N *coefficients* of the polynomials of the *next* generation identified with the N *zeros* of the polynomials of the *current* generation—and to the related possibility to identify *endless hierarchies* of *nonlinear solvable* dynamical systems, including, again, N-body problems describing the evolution in the *complex* plane (or, equivalently, in the *real* Cartesian plane) of N point-particles interacting *nonlinearly* among themselves and moving according to *Newtonian* equations of motion ("accelerations equal forces").

Some other developments that have also followed from the key formulas reported in Chapter 2 are treated in other chapters and sections of this book: the interested reader may immediately identify them from the titles of the relevant chapters, sections and subsections, as displayed in the Table of Contents.

Let us end this terse Chapter 1 by displaying, without any immediate explanation, 2 *rather trivial examples* of the type of N-body problems—for simplicity, with $N = 2$—the *solvability* of which, demonstrated in this book, entails an understanding of their *remarkable* phenomenology. Hence, in the rest of this short Chapter 1 we refer to 2 point particles moving in the *real* Cartesian plane, while in the rest of this book we mainly focus on N-body problems—with $N \geq 2$—characterizing the motion of N points moving in the *complex* plane.

Example 1.1. The equations of motion of the *first* of these 2 *solvable* models read as follows:

$$\ddot{\vec{r}}_n = (-1)^{n+1} \frac{2}{r^2} \left[\left(\dot{\vec{r}}_1 \cdot \vec{r} \right) \dot{\vec{r}}_2 + \left(\dot{\vec{r}}_2 \cdot \vec{r} \right) \dot{\vec{r}}_1 - \left(\dot{\vec{r}}_1 \cdot \dot{\vec{r}} \right) \vec{r} \right]$$

$$- (\rho_1 \, \omega)^2 \, \vec{r}_n + \left[2 \, (\rho_1)^2 - (\rho_2)^2 \right] \omega^2 \, r^{-2} \left[(r_{n+1})^2 \, \vec{r}_n - (r_n)^2 \, \vec{r}_{n+1} \right],$$

$$\vec{r} \equiv \vec{r}_1 - \vec{r}_2, \quad n = 1, 2 \mod [2]. \tag{1.1}$$

Here the 2-vectors $\vec{r}_1 \equiv \vec{r}_1(t)$ and $\vec{r}_2 \equiv \vec{r}_2(t)$ are the coordinates of 2 unit-mass point-particles moving in the *real* Cartesian plane as functions of the *independent* variable t ("time"); superimposed dots indicate time differentiations; the symbol $\cdot$ denotes the standard *scalar product* in the plane (so that, in particular, $r^2 = \vec{r} \cdot \vec{r}$ is the squared modulus of the 2-vector $\vec{r} \equiv \vec{r}_1 - \vec{r}_2$); and ρ_1, ρ_2, and ω are 3 *arbitrary real nonvanishing* parameters. It is shown below that this model is *Hamiltonian* and solvable by *algebraic* operations: in fact by just finding the 2 zeros of a second-degree time-dependent polynomial explicitly known in terms of the initial data of the problem, $\vec{r}_n(0)$ and $\dot{\vec{r}}_n(0)$. Moreover—if the 2 parameters ρ_1, ρ_2 are both *rational* numbers—then this model is *isochronous*, its *generic* solutions being *all* periodic with a common period which is an *integer* multiple of the basic period $T = 2\pi / |\omega|$. Note that this model is invariant under a common multiplicative rescaling of the dependent variables $\vec{r}_n \equiv \vec{r}_n(t)$, hence it is in particular *invariant under rotations* of the Cartesian plane (as demonstrated by the covariance of its equations of motion (1.1)); while it is *not* translation-invariant. ∎

Example 1.2. The equations of motion of the *second* of these 2 *solvable* models read as follows:

$$\ddot{\vec{X}} = \omega \, \hat{\vec{X}} + \rho_1 \, \omega^2 \, \vec{X} + \frac{1}{X^2} \left[2 \left(\dot{\vec{X}} \cdot \vec{X} \right) \dot{\vec{X}} - \left(\dot{\vec{X}} \cdot \dot{\vec{X}} \right) \vec{X} \right], \tag{1.2a}$$

$$\ddot{\vec{x}} = -\frac{1}{2} \, (\rho_2 \, \omega)^2 \, \vec{x} - \frac{1}{x^2} \left\{ 2 \left(\dot{\vec{x}} \cdot \vec{x} \right) \dot{\vec{x}} - \left(\dot{\vec{x}} \cdot \dot{\vec{x}} \right) \vec{x} \right.$$

$$\left. + \left[\rho_1 + (\rho_2)^2 / 2 \right] \omega^2 \left[2 \left(\vec{X} \cdot \vec{x} \right) \vec{X} - X^2 \, \vec{x} \right] \right.$$

$$+ \omega \left[\left(\dot{\vec{X}} \cdot \vec{x} \right) \hat{X} + \left(\vec{X} \cdot \vec{x} \right) \dot{\hat{X}} - \left(\dot{\vec{X}} \cdot \vec{X} \right) \hat{x} \right]$$

$$+ \left[2 \left(\dot{\vec{X}} \cdot \vec{x} \right) \dot{\vec{X}} + \left(\dot{\vec{X}} \cdot \dot{\vec{X}} \right) \vec{x} \right] \Big\} . \tag{1.2b}$$

Here the 2 *dependent* variables $\vec{X} \equiv \vec{X}(t) \equiv (X_1(t), X_2(t))$ and $\vec{x} \equiv \vec{x}(t) \equiv (x_1(t), x_2(t))$ are 2-vectors describing the motion in the *real* Cartesian plane of 2 unit-mass point-particles as functions of the *independent* variable t ("time"); $\hat{X} \equiv \hat{X}(t) \equiv (-X_2(t), X_1(t))$, and $\hat{x} \equiv \hat{x}(t) \equiv (-x_2(t), x_1(t))$ are these vectors rotated in the Cartesian plane by $\pi/2$; $X \equiv X(t)$ and $x \equiv x(t)$ are the moduli of these 2-vectors; the symbol $\cdot$ denotes the standard *scalar product* in the plane (so that, for instance, $x^2 = \vec{x} \cdot \vec{x} = \hat{x} \cdot \hat{x}$); superimposed dots indicate time-differentiations; and the 3 parameters ρ_1, ρ_2, ω are 3 *arbitrary real* parameters. Note that we used the *unambiguous* notation $\dot{\vec{X}} \cdot \dot{\vec{X}} \equiv \left(\dot{X}_1 \right)^2 + \left(\dot{X}_2 \right)^2$ (instead of the *ambiguous* notation $\left(\dot{X} \right)^2$ that may rather be interpreted to mean $\left\{ d/dt \left[(X_1)^2 + (X_2)^2 \right]^{1/2} \right\}^2 = \left[\left(\dot{X}_1 X_1 + \dot{X}_2 X_2 \right) /X \right]^2$); and likewise for $\dot{\vec{x}} \cdot \dot{\vec{x}} \equiv (\dot{x}_1)^2 + (\dot{x}_2)^2$. Also note that these equations of motion are *scale-invariant*, i.e., *invariant* under the *rescaling* transformation $X(t) \Rightarrow c\,X(t)$, $x(t) \Rightarrow c\,x(t)$ with c an *arbitrary nonvanishing constant*; hence, they are *covariant*, i.e., *rotation-invariant* in the plane. And they are clearly *Newtonian*: accelerations equal forces, with the forces depending *nonlinearly* on the coordinates of the particles and on their velocities.

A *remarkable* feature of this system of 2 equations of motion (see (1.2)) is that—if the 2 parameters ρ_1 and ρ_2 are *rational* numbers,

$$\rho_n = \frac{q_n}{k_n} \tag{1.2c}$$

where q_n and k_n are two *arbitrary coprime integers* ($q_n \neq 0$, $k_n \geq 1$, $n = 1, 2$), and ω is an *arbitrary nonvanishing real* parameter to which the period $T = 2\pi/|\omega|$ is associated—then this model is *isochronous*: there is a *large open* set of initial data $\vec{x}(0), \dot{\vec{x}}(0), \vec{X}(0), \dot{\vec{X}}(0)$ yielding *nonsingular* evolutions that are *all periodic* with a period that is an *integer* multiple of the basic period $T = 2\pi/|\omega|$. ∎

These findings are merely *rather trivial quite special* examples of those discussed later in the book (see Chapter 4, where however we generally consider *N*-body problems with *arbitrary* $N \geq 2$). The interested reader will find their proofs there, and thereby learn how to obtain the *explicit* solutions of the initial-value problems of these 2 Newtonian 2-body systems in the plane.

2

Parameter-Dependent Monic Polynomials

Definitions, Key Formulas and Other Preliminaries

In Chapter 2 we detail our notation for parameter-dependent monic polynomials and we then report the formulas relating the derivatives—with respect to the parameter—of the *zeros* of such a polynomial to the analogous derivatives of the *coefficients* of this polynomial. Most of the results reported in this book are based on these key formulas. Moreover, we will discuss other results that are also fundamental for the rest of this book: (i) the *periodicity* properties of the *zeros* of a time-dependent polynomial that is itself, *periodic*; (ii) a convenient transformation applicable to a class of nonlinear evolutionary Ordinary Differential Equations (ODEs) which causes the *transformed* ODEs thereby obtained to feature the remarkable property to be *isochronous*, i.e. to possess an *open* set of *initial* data yielding solutions *all* of which are *periodic* with a *fixed* period (independent of the initial data inside that *open* set, which is generally *quite large*); (iii) specific examples of such ODEs.

2.1 Notation

In Section 2.1 we introduce the notation for parameter-dependent polynomials that will be used throughout this book.

A monic polynomial of degree N in its argument z that is dependent on a parameter t is hereafter denoted as follows:

$$p_N(z; \vec{y}(t), \tilde{x}(t)) = z^N + \sum_{m=1}^{N} \left[y_m(t) \, z^{N-m} \right] = \prod_{n=1}^{N} [z - x_n(t)] . \qquad (2.3)$$

Hereafter—unless otherwise indicated—N is an *arbitrary positive integer* (generally, $N \geq 2$), *indices* such as n, m run over the *positive integers* from 1 to N, t is a *real* parameter generally having the significance of "time", the N components $y_m(t)$ of the N-vector $\vec{y}(t)$ are the N *coefficients* of the polynomial $p_N(z; \vec{y}(t), \tilde{x}(t))$, and the N elements of the *unordered* set $\tilde{x}(t)$ are the N *zeros* of this polynomial. Note that often, for notational convenience, the t dependence of the various quantities will *not* be explicitly displayed; for instance, we will write simply y_m or x_n rather than $y_m(t)$ or $x_n(t)$. Note that the notation $p_N(z; \vec{y}, \tilde{x})$, see (2.3), is *redundant*: this polynomial is completely identified by assigning *either* its N coefficients y_m *or* its N zeros x_n. Indeed the N coefficients y_m of the polynomial $p_N(z; \vec{y}, \tilde{x})$ are *explicitly* expressed by well-known formulas in terms of the N *symmetrical* sums $\sigma_m(\tilde{x})$ of the N zeros x_n of this polynomial,

$$y_m = (-1)^m \, \sigma_m(\tilde{x}) \,, \tag{2.4a}$$

$$\sigma_m(\tilde{x}) = \sum_{1 \le n_1 < n_2 < \cdots < n_m \le N} \left(x_{n_1} \cdot x_{n_2} \cdots x_{n_m} \right)$$

$$= \frac{1}{m!} \sum_{n_1, n_2, \ldots, n_m = 1}^{N} \left(x_{n_1} \cdot x_{n_2} \cdots x_{n_m} \right). \tag{2.4b}$$

Likewise the N zeros x_n are uniquely identified—up to their $N!$ permutations—once the N coefficients y_m of the polynomial $p_N(z; \vec{y}, \tilde{x})$ are assigned, although *explicit* formulas expressing the *zeros* x_n in terms of the *coefficients* y_m are generally *only* available—in terms of elementary functions such as quadratic, cubic respectively quartic roots—for $N = 2, N = 3$ respectively $N = 4$.

Hereafter—unless otherwise indicated—we assume the quantities z, y_m and x_n to be *complex* numbers; and we generally focus—unless otherwise specified—on *generic* polynomials characterized by *generic* values of their N *complex coefficients* and N *complex zeros*—with the N zeros being *all different among themselves*. And we adopt hereafter the standard convention, which states that *empty sums vanish* and *empty products equal unity*, $\sum_{\ell=L}^{L'} (\cdot) = 0$ and $\prod_{\ell=L}^{L'} (\cdot) = 1$ if $L > L'$.

For future reference, we also report the *explicit* expressions of the *t*-derivatives of the coefficients $y_m(t)$, which clearly follow from the formulas (2.4):

$$\dot{y}_1 = -\sum_{n=1}^{N} \dot{x}_n \,, \tag{2.5a}$$

$$\dot{y}_m = (-1)^m \sum_{j=1}^{m} \left[\dot{x}_j \sum_{1 \le n_1 < n_2 < \cdots < n_{j-1} < n_{j+1} < \cdots < n_m \le N;} \left(x_{n_1} x_{n_2} \cdots x_{n_m} \right) \right],$$

$$m = 2, \ldots, N \,, \tag{2.5b}$$

as well as the obvious identities

$$\sum_{m=1}^{N} \left[y_m (x_n)^{N-m} \right] = -(x_n)^N \,. \tag{2.6}$$

Remark 2.1.1. As emphasized by the notation $\tilde{x}$ denoting the *unordered* set of N (generally *complex*) numbers x_n, the identification of the N zeros x_n—namely, the association of the (generally *complex*) value of the zero x_n to its index n—is *ambiguous* since the N zeros x_n are only defined up to $N!$ permutations of their indices. But—in the case of polynomials *continuously* dependent on a parameter t (say, "continuous time"—as those generally considered throughout this book, except in Chapter 7)—this *ambiguity* is obviously restricted *only* to the N values of the *zeros* at a specific value of the *(real)* parameter t: say, to the "initial" values $x_n(0)$ of the N *zeros* at $t = 0$. Indeed if the N *zeros* of the polynomial $p_N(z; \vec{y}(t), \tilde{x}(t))$ feature a *continuous* dependence on the parameter t throughout their t-evolution from $t = 0$—and moreover do *not* collide, thereby *losing* their identities—then the identity of each of them (as specified by the value of the index n associated to each of them) is determined for *all* values of the parameter t by *continuity* in t from their identity at $t = 0$.

However, when the technique of solution of a system of ODEs that characterizes the evolution of N points $x_n(t)$ moving in the *complex* x-plane is based on their identification as the N *zeros* of a t-dependent polynomial of degree N—the main tool used throughout this book—then this technique of solution provides the *configuration* of the system at time t as an *unordered* set of N coordinates $x_n(t)$ but it does *not* allow to identify, say, which is the value at time t of the specific coordinate $x_1(t)$ that has evolved by *continuity* in t from the initial data (say, $x_1(0)$ and $\dot{x}_1(0)$ for second-order ODEs). One way to do so is by tracing the time evolution of each coordinate over the *Riemann surface* associated with the configuration of the *zeros* of the polynomial (2.3). This is not a trivial endeavor, as demonstrated by various papers where this phenomenology has been studied in considerable detail [76], [89], [55], [77], [56], [75]. Another, more practical way is to integrate *numerically* the equations of motion from the initial data (possibly only with rather poor precision); or to chop up the time interval from 0 to t into several (say, s) subintervals (from 0 to t_1, from t_1 to t_2, ..., from t_{s-1} to $t_s = t$), to solve in every subinterval by the technique described below, and to make sure that each subinterval is sufficiently short to allow the identification of each moving point by an argument of *contiguity* (approximating *continuity*) of their positions over their time evolution.

On the other hand, often the technique of solution described below also yields some important *general information* on the behavior of the solutions of the system under consideration, such as their *periodicity properties*: see below. ∎

Let us end this section by emphasizing the relevance of this Remark 2.1.1 to most of the following developments.

2.2 Key Formulas

The formulas relating the t-derivatives of the N *zeros* $x_n(t)$ of the t-dependent monic polynomial $p_N(z; \vec{y}(t), \tilde{x}(t))$, see (2.3), to the analogous t-derivatives of the

N coefficients $y_m(t)$ of this polynomial underline essentially *all* the results reported
in this book. The formulas for the derivatives of order 1 and 2—which play a major
role in the following—are displayed in Subsection 2.2.1 and proven in Subsection
2.2.3; some of those for derivatives of higher order are reported in Subsection 2.2.2,
and indications of where their proofs can be found are provided in Section 2.N. Let
us emphasize that these formulas are *identities* valid for any parameter-dependent
generic polynomial; they should be replaced by their limiting forms in the case of
nongeneric polynomials featuring *multiple zeros*.

Hereafter superimposed dots denote *t*-derivatives, for instance $\dot{x}_n(t) \equiv d\,x_n(t)/dt$,
$\ddot{y}_m(t) \equiv d^2 y_m(t)/dt^2$.

2.2.1 First-Order and Second-Order t-Derivatives of the Zeros
of a Monic t-Dependent Polynomial

The following formula relates the *first-order* *t*-derivative $\dot{x}_n(t)$ of the *zero* $x_n(t)$ of
a generic *t*-dependent polynomial $p_N(z; \vec{y}(t), \tilde{x}(t))$, see (2.3), to the *N first-order*
t-derivatives $\dot{y}_m(t)$ of the *N* coefficients $y_m(t)$ of the same polynomial:

$$\dot{x}_n = -\left[\prod_{\ell=1,\,\ell\neq n}^{N} (x_n - x_\ell) \right]^{-1} \sum_{m=1}^{N} \left[\dot{y}_m \, (x_n)^{N-m} \right]. \tag{2.7}$$

The analogous formula for *second-order* derivatives reads as follows:

$$\ddot{x}_n = \sum_{\ell=1,\,\ell\neq n}^{N} \left(\frac{2\,\dot{x}_n\,\dot{x}_\ell}{x_n - x_\ell} \right) - \left[\prod_{\ell=1,\,\ell\neq n}^{N} (x_n - x_\ell) \right]^{-1} \sum_{m=1}^{N} \left[\ddot{y}_m \, (x_n)^{N-m} \right]. \tag{2.8}$$

The inverse formula expressing the *t*-derivatives of the coefficients $y_m(t)$ in terms
of the zeros $x_n(t)$ and their *t*-derivatives, see (2.5), is an immediate consequence of
(2.4). We do not report the analogous formula for second derivatives since we do
not use it in the following.

2.2.2 Third-Order and Fourth-Order t-Derivatives of the Zeros
of a Monic t-Dependent Polynomial

In Subsection 2.2.2 we display the formulas relating the *third-order* and *fourth-
order* *t*-derivatives of the *zero* $x_n(t)$ of a generic *t*-dependent polynomial
$p_N(z; \vec{y}(t), \tilde{x}(t))$, see (2.3), to the corresponding *t*-derivatives of the *N* coefficients
$y_m(t)$ of the same polynomial:

$$\dddot{x}_n = 3 \sum_{\ell=1;\ \ell\neq n}^{N} \left(\frac{\ddot{x}_n\ \dot{x}_\ell + \ddot{x}_\ell\ \dot{x}_n}{x_n - x_\ell} \right)$$

$$- 3 \sum_{\ell_1,\ell_2=1;\ \ell_1\neq n,\ \ell_2\neq n}^{N} \left[\frac{\dot{x}_n\ \dot{x}_{\ell_1}\ \dot{x}_{\ell_2}}{(x_n - x_{\ell_1})\ (x_n - x_{\ell_2})} \right]$$

$$- \left[\prod_{\ell=1,\ \ell\neq n}^{N} (x_n - x_\ell)^{-1} \right] \sum_{m=1}^{N} \left[\dddot{y}_m\ (x_n)^{N-m} \right], \tag{2.9}$$

$$\ddddot{x}_n = \sum_{\ell=1}^{N} \left(\frac{4\ \dddot{x}_n\ \dot{x}_\ell + 4\ \dddot{x}_\ell\ \dot{x}_n + 6\ \ddot{x}_n\ \ddot{x}_\ell}{x_n - x_\ell} \right)$$

$$- 6 \sum_{\ell_1,\ell_2=1;\ \ell_1\neq n,\ \ell_2\neq n}^{N} \left[\frac{\ddot{x}_n\ \dot{x}_{\ell_1}\ \dot{x}_{\ell_2} + 2\ \ddot{x}_{\ell_1}\ \dot{x}_{\ell_2}\ \dot{x}_n}{(x_n - x_{\ell_1})\ (x_n - x_{\ell_2})} \right]$$

$$+ 4 \sum_{\ell_1,\ell_2,\ell_3=1;\ \ell_1\neq n,\ \ell_2\neq n,\ \ell_2\neq n}^{N} \left[\frac{\dot{x}_n\ \dot{x}_{\ell_1}\ \dot{x}_{\ell_2}\ \dot{x}_{\ell_3}}{(x_n - x_{\ell_1})\ (x_n - x_{\ell_2})\ (x_n - x_{\ell_3})} \right]$$

$$- \left[\prod_{\ell=1,\ \ell\neq n}^{N} (x_n - x_\ell)^{-1} \right] \sum_{m=1}^{N} \left[\ddddot{y}_m\ (x_n)^{N-m} \right]. \tag{2.10}$$

2.2.3 Proofs

In Subsection 2.2.3 we report the proofs of (2.7) and (2.8).

The starting point to prove the relation (2.7) are the two formulas

$$\left(\frac{d}{dt} \right) p_N\left(z; \vec{y}(t), \tilde{x}(t)\right) = \sum_{m-1}^{N} \left[\dot{y}_m\ z^{N-m} \right], \tag{2.11a}$$

$$\left(\frac{d}{dt} \right) p_N\left(z; \vec{y}(t), \tilde{x}(t)\right) = - \sum_{n=1}^{N} \left[\dot{x}_n \prod_{\ell=1,\ \ell\neq n}^{N} (z - x_\ell) \right], \tag{2.11b}$$

which clearly obtain by *t*-differentiation of the expressions of the polynomial $p_N(z; \vec{y}(t), \tilde{x}(t))$ in terms of its *N coefficients* $y_m(t)$, respectively of its *N zeros* $x_n(t)$, see (2.3). They imply the relation

$$\sum_{n=1}^{N} \left[\dot{x}_n \prod_{\ell=1,\ \ell\neq n}^{N} (z - x_\ell) \right] = - \sum_{m=1}^{N} \left[\dot{y}_m\ z^{N-m} \right], \tag{2.11c}$$

and, for $z = x_n$, this formula yields (2.7), which is thereby proven.

Likewise, an additional *t*-differentiation of (2.11a) yields

$$\left(\frac{d}{dt}\right)^2 p_N\left(z; \vec{y}\,(t)\,, \tilde{x}\,(t)\right) = \sum_{m=1}^{N} \left(\ddot{y}_m\, z^{N-m}\right)\,, \tag{2.12a}$$

while an additional *t*-differentiation of (2.11b) yields

$$\left(\frac{d}{dt}\right)^2 p_N\left(z; \vec{y}\,(t)\,, \tilde{x}\,(t)\right) = -\sum_{n=1}^{N}\left\{\ddot{x}_n\left[\prod_{\ell=1,\,\ell\neq n}^{N}(z - x_\ell)\right]\right\}$$

$$+ \sum_{\ell_1,\ell_2=1,\,\ell_1\neq\ell_2}^{N}\left\{\dot{x}_{\ell_1}\,\dot{x}_{\ell_2}\left[\prod_{\ell'=1,\,\ell'\neq\ell_1,\ell_2}^{N}(z - x_{\ell'})\right]\right\}$$

$$= \sum_{m=1}^{N}\left(\ddot{y}_m\, z^{N-m}\right)\,, \tag{2.12b}$$

where the second equality is implied by (2.12a). Hence, for $z = x_n$, one gets (2.8), which is thereby proven.

2.3 Periodicity of the Zeros of a Time-Dependent Periodic Polynomial

In Section 2.3 we discuss *tersely* the topic indicated in its title, and we provide a key reference for a more detailed treatment of this topic.

Suppose that a time-dependent polynomial $p_N(z; t)$ such as (2.3)—which we assume in this section to *evolve continuously* over time and to be *generic* for *all* time, i.e., such that its N zeros $x_n(t)$ are for *all* time all different among themselves, $x_n(t) \neq x_\ell(t)$ if $n \neq \ell$—is *completely periodic* with period T, i.e., for *all* values of z:

$$p_N\left(z; t+T\right) = p_N\left(z;\, \vec{y}\,(t+T)\,, \tilde{x}\,(t+T)\right) = p_N\left(z;\, t\right) = p_N\left(z;\, \vec{y}(t),\, \tilde{x}(t)\right)\,. \tag{2.13a}$$

This obviously implies that its N *coefficients* are as well *periodic* with the *same* period T:

$$\vec{y}\,(t+T) = \vec{y}\,(t)\,, \quad y_m\,(t+T) = y_m\,(t)\,, \tag{2.13b}$$

and this is also clearly true for the *unordered* set of its N zeros:

$$\tilde{x}(t+T) = \tilde{x}\,(t)\,. \tag{2.13c}$$

But this need not be true for *each* of its N zeros $x_n(t)$, which we assume to evolve themselves *continuously* over time; so that the *unordered* character of the set $\tilde{x}(t)$ is only relevant, say, at the initial time $t = 0$, because the assignment of its label n to the *zero* x_n can only be *arbitrarily assigned*—say, at the initial time $t = 0$—remaining thereafter *determined* by the assumed *continuity* of $x_n(t)$ as a function of time t (see Remark 2.1.1). So the last formula—due to the possibility that, as it were, some *zeros* exchange their roles after one period T— does *not* imply $x_n(t + T) = x_n(t)$ but only

$$x_n(t + \nu T) = x_n(t) , \tag{2.14a}$$

with ν a *positive integer* obviously *not larger* than N!, $\nu \leq N!$ ($N!$ being the *maximal* number of *different* permutations of N items). But in fact—because of the *dual* nature of the exchange of roles among *zeros*—the *maximal* value $\nu_{\text{Max}}(N)$ of ν,

$$\nu \leq \nu_{\text{Max}}(N) , \tag{2.14b}$$

is *much smaller* than $N!$ (for large N). The reader interested in a more detailed discussion of this topic—including a table of *all possible values* of ν for N up to 12 and *asymptotic estimates* of the number $\nu_{\text{Max}}(N)$ for large N—is advised to study the *clear* treatment of this question in [76] (and for a detailed discussion of this phenomenology in analogous many-body contexts see [55, 89, 56, 75]).

> **Remark 2.3.1.** Clearly the regions of initial data $\tilde{x}(0)$ yielding time evolutions with *different* periods are *separated* from each other by boundaries out of which emerge solutions such that the evolution equations—at some time t_s such that $0 < t_s < T$—feature a *singularity* due to a collision of two (or exceptionally more) different zeros, $x_n(t_s) = x_\ell(t_s), n \neq \ell$. ∎

> **Remark 2.3.2.** Essentially everything that has been written thus far about the *periodicity* of the N zeros $x_n(t)$ of a time-dependent polynomial $p_N(z; t)$ which is itself *periodic*, is equally valid if the *periodicity* property—see (2.13)—is replaced by the property of *asymptotic periodicity*, i.e.,
>
> $$\lim_{t \to \infty} \left[p_N(z; t + T) - p_N(z; t) \right] = 0 . \blacksquare \tag{2.15}$$

2.4 How Certain Evolution Equations Can Be Made Periodic

In Section 2.4 we tersely review a *trick* that allows to modify certain evolution equations so that the modified equations thereby obtained feature many, or perhaps only or almost only, *periodic* solutions. For simplicity we illustrate this *trick* in the simple context of a *single scalar autonomous* ODE of *second* order, but the alert reader shall immediately understand how this finding can be extended to more general contexts: for instance, ODEs of *different* orders, *systems* of such equations, or possibly *nonautonomous* equations. These results provide the basis for many of the findings reported in Chapter 4.

Let us therefore focus on the following (*autonomous*) ODE:

$$\gamma'' = f(\gamma', \gamma; r) , \tag{2.16a}$$

where $\gamma \equiv \gamma(\tau)$ is the dependent variable, τ is the independent variable, appended primes denote differentiations with respect to τ, and there is a *real rational* value of the parameter r,

$$r = \frac{q}{k} \tag{2.16b}$$

with q and k two *coprime integers* and, for definiteness, $k \geq 1$, for which the function $f(w, v; r)$ satisfies the following *scaling* property:

$$f\left(a^{1+r}\, w,\, a^r\, v;\, r\right) = a^{2+r} f\left(w,\, v;\, r\right) , \tag{2.16c}$$

with a an *arbitrary* parameter. Hereafter we assume to always work with *complex* numbers, unless otherwise indicated.

The typical example of function $f(w, v;\ r)$ satisfying the condition (2.16c) is

$$f\left(w,\, v;\ r\right) = c\, w^{\alpha}\, v^{1-\alpha+(2-\alpha)/r} , \tag{2.17}$$

where we generally assume the parameter c to be an arbitrary *complex* number and the parameter α, as the parameter r, to be a *real rational* number—although this last condition is not actually necessary in order that (2.17) satisfy (2.16c)). Indeed, the diligent reader will have no difficulty to verify that this function $f(w, v;\ r)$ satisfies the condition (2.16c)—and this would remain true if in the right-hand side of (2.17) there appeared a sum of an *arbitrary* number of other analogous terms with different assignments of the parameters c and α (but the same parameter r).

Let us now perform a change of dependent and independent variables on the ODE (2.16) by introducing a new function of the *real* variable t ("time") via the position

$$y(t) = \exp\left(\mathbf{i}\, r\, \omega\, t\right) \gamma(\tau) , \quad \tau \equiv \tau(t) = \frac{\exp\left(\mathbf{i}\, \omega\, t\right) - 1}{\mathbf{i}\, \omega} , \tag{2.18a}$$

where (here and throughout) $\mathbf{i}$ is the *imaginary unit* (so that $\mathbf{i}^2 = -1$) and ω is an *arbitrary real nonvanishing* parameter to which we associate the basic period

$$T = \frac{2\pi}{|\omega|} . \tag{2.18b}$$

It is then easily seen that the ODE (2.16) implies that the function $y(t)$ satisfies—thanks to the *scaling* property (2.16c)—the *autonomous* ODE

$$\ddot{y} = (2\,r + 1)\, \mathbf{i}\, \omega\, \dot{y} + r\,(r + 1)\, \omega^2\, y + f\,(\dot{y} - \mathbf{i}\, r\, \omega\, y,\, y;\, r) . \tag{2.19a}$$

It is then clear—see (2.18)—that this ODE (2.19a) is likely to possess lots of periodic solutions. Indeed *all* its solutions $y(t)$ which correspond via (2.18) to solutions $\gamma(\tau)$ of the ODE (2.16) which are themselves *entire* functions of the *complex* variable τ are clearly *periodic* with period kT (see (2.16b) and (2.18)):

$$y(t + k\,T) = y(t)\ .\tag{2.19b}$$

And solutions $\gamma(\tau)$ which are *not entire* functions of τ but have a *simple analytic* dependence on the variable τ shall also entail *simple periodicity* properties for the corresponding solutions of the ODE (2.19a). In the following subsections of this Section 2.4 we identify several *solvable* ODEs of type (2.16) and discuss the *periodicity* properties of their variant (2.19a); these findings shall play an important role below.

Remark 2.4.1. It is also of interest for future developments to replace the assumption made above that ω is an arbitrary *real* nonvanishing parameter with the hypothesis that ω is instead an arbitrary *complex* nonvanishing parameter with, say, *positive imaginary part*, Im$[\omega] > 0$. This possibility will be tersely explored at the end of each of the following subsections. ∎

Remark 2.4.2. In the following subsections of this Section 2.4, we discuss tersely several examples of ODEs belonging to the class (2.16) that are *explicitly solvable* in terms of *elementary* functions. In these treatments we forsake the possibility to introduce additional *free* parameters in these equations via the trick to rescale and shift the *dependent* variable—although this trivial trick might play a less trivial role below, by increasing the number of *free* parameters featured by the models under consideration. ∎

Let us finally mention that at this point some readers might well prefer to skip the following subsections of this Section 2.4 and only return to them when their findings are referred to below (mainly in the later sections of Chapter 4).

2.4.1 The ODE $\gamma'' = c\,(\gamma')^2\,\gamma^{-1}$ and Its Isochronous Variant

In Subsection 2.4.1 the solution of the ODE

$$\gamma'' = c\,(\gamma')^2\,\gamma^{-1}\tag{2.20}$$

is reported. Here $\gamma \equiv \gamma(\tau)$, primes denote differentiations with respect to the independent variable τ, and c is an *a priori arbitrary complex* number (but see below for special values of this parameter).

Remark 2.4.1.1. This equation of motion, (2.20), is *Hamiltonian*, being yielded by the Hamiltonian

$$h(p;\,\gamma) = a\,[p\,\gamma^c]^\beta\ ,\tag{2.21}$$

where, in the Hamiltonian context, the variable $p \equiv p(\tau)$ plays the role of *canonical momentum* conjugated to the *canonical variable* $\gamma \equiv \gamma(\tau)$, and the (*nonvanishing*) parameters a and β may be *arbitrarily assigned* ($a \neq 0$, $\beta \neq 0$, $\beta \neq 1$). ∎

The modified variant (see (2.19a)) of this equation (2.20) reads as follows:

$$\ddot{y} = [2\,r\,(1-c) + 1]\,\mathbf{i}\,\omega\,\dot{y} + r\,[r\,(1-c)+1]\,\omega^2\,y + c\,\frac{(\dot{y})^2}{y}. \tag{2.22}$$

It features the *real* independent variable t ("time"), $y \equiv y(t)$ (here superimposed dots denote time-differentiations). Let us discuss it, with particular attention to the *periodicity* properties of its solutions. Note that the ODE (2.20) corresponds to (2.17) with $\alpha = 2$ and r essentially *arbitrary*, hence hereafter (in Subsection 2.4.1) r is an *arbitrary real rational* number, see (2.16b). Of course in the following discussion of the ODEs (2.20) we exclude from consideration the initial datum $\gamma(0) = 0$, and likewise we exclude the initial datum $y(0) = 0$ in the following discussion of the ODE (2.22).

It is easily seen that the solution of the initial-value problem of the ODE (2.20) reads as follows: For $c \neq 1$:

$$\gamma(\tau) = \gamma(0)\,\left(1 - \frac{\tau}{\bar{\tau}}\right)^{1/(1-c)}, \quad \bar{\tau} = \frac{\gamma(0)}{(c-1)\,\gamma'(0)}; \tag{2.23a}$$

for $c = 1$

$$\gamma(\tau) = \gamma(0)\,\exp\left[\frac{\gamma'(0)\,\tau}{\gamma(0)}\right]. \tag{2.23b}$$

Hence the solution of the initial-value problem of the ODE (2.22) reads as follows: for $c \neq 1$

$$y(t) = \tilde{y}(\eta), \quad \tilde{y}(\eta) = y(0)\,\eta^r\,\left(\frac{\bar{\eta} - \eta}{\bar{\eta} - 1}\right)^{1/(1-c)},$$

$$\eta \equiv \eta(t) = \exp(\mathbf{i}\,\omega\,t), \quad \bar{\eta} = 1 + \frac{\mathbf{i}\,\omega\,y(0)}{(c-1)\,[\dot{y}(0) - \mathbf{i}\,r\,\omega\,y(0)]}; \tag{2.24a}$$

for $c = 1$

$$y(t) = \tilde{y}(\eta), \quad \eta \equiv \eta(t) = \exp(\mathbf{i}\,\omega\,t),$$

$$\tilde{y}(\eta) = y(0)\,\eta^r\,\exp\left\{\left[\frac{\dot{y}(0)}{\mathbf{i}\,\omega\,y(0)} - r\right]\,(\eta - 1)\right\}. \tag{2.24b}$$

These formulas imply that, for $c = 1$, the ODE (2.22)—which, in this case, reads simply

$$\ddot{y} = \mathbf{i}\,\omega\,\dot{y} + r\,\omega^2\,y + \frac{(\dot{y})^2}{y} \tag{2.25}$$

—is *isochronous: all* its solutions are *nonsingular* and *periodic* with period kT, see (2.19b), (2.18b) and (2.16b).

The situation in the $c \neq 1$ case is more nuanced. There clearly still is an *open* set of initial data—characterized by the *inequality* $|\bar{\eta}| > 1$, i.e.,

$$|\bar{\eta}| = \left| 1 + \frac{\mathbf{i}\,\omega\,y\,(0)}{(c-1)\,\left[\dot{y}\,(0) - \mathbf{i}\,r\,\omega\,y\,(0)\right]} \right| > 1 \qquad (2.26)$$

—such that, again, *all* the solutions of the ODE (2.22) are *nonsingular* and *periodic* with period kT, see (2.19b), (2.18b) and (2.16b) (and note, incidentally, that this *inequality* is automatically satisfied for arbitrary $y(0) \neq 0$ if $\dot{y}(0) = 0$ and $c = 1+\mathbf{i}R$ with R an *arbitrary nonvanishing real* number). On the other hand, the solutions (2.24a) characterized by the *complementary inequality* $|\bar{\eta}| < 1$, i.e.,

$$|\bar{\eta}| = \left| 1 + \frac{\mathbf{i}\,\omega\,y\,(0)}{(c-1)\,\left[\dot{y}\,(0) - \mathbf{i}\,r\,\omega\,y\,(0)\right]} \right| < 1 \,, \qquad (2.27a)$$

while also *nonsingular* for all time, are *not periodic* because, as a function of real t, the *complex* number $\tilde{y}(\eta) \equiv \tilde{y}\,(\eta\,(t))$, see (2.24a), travels on an infinitely-sheeted Riemann surface; unless the parameter c is itself a *real rational* number,

$$c = \frac{q_c}{k_c} \quad \text{implying} \quad \frac{1}{1-c} = \frac{k_c}{k_c - q_c} \qquad (2.27b)$$

with q_c and k_c two *different coprime integers* (*a priori arbitrary*, but for definiteness $k_c \geq 1$), in which case the function $\tilde{y}(\eta)$ of the *complex* variable η features a *rational* branch point at $\eta = \bar{\eta}$ (see (2.24a)). Hence clearly these solutions (2.24a) are *again periodic*, but now with, at most, period KT,

$$y\,(t + K\,T) = y\,(t) \,, \qquad (2.27c)$$

where (see (2.16b) and (2.27b))

$$K = \text{MinimumCommonMultiple}\,\left[k, \ |k_c - q_c|\right] \,. \qquad (2.28)$$

Remark 2.4.1.2. Here—and occasionally below—we assert that a solution is periodic with, "at most", a certain period. This means that we leave to the interested reader the, generally easy, task to ascertain more precisely what the exact period in question might be. For instance, the function $a\cos\left[(6/5)\,\omega t\right] + b\sin\left[(3/7)\,\omega t\right]$ is periodic in t with "at most" period $5 \cdot 7 \cdot T = 35\,T$ with $T = 2\pi/\,|\omega|$, but—more precisely—it is periodic with the smaller period $(35/3)\,T$—which of course implies that it is also periodic with the larger period $35\,T$. ∎

There remains to consider, in the $c \neq 1$ case, what happens in the *nongeneric* case when the initial data satisfy the *equality*

$$|\bar{\eta}| = \left|1 + \frac{\mathbf{i}\,\omega\,y(0)}{(c-1)\,\left[\dot{y}(0) - \mathbf{i}\,r\,\omega\,y(0)\right]}\right| = 1\,. \tag{2.29}$$

Then, the solution (2.24a) of the ODE (2.22) features a *singularity* at the (*real*) times $t = t_s$, such that $\exp(\mathbf{i}\,\omega\,t_s) = \bar{\eta}$ (see (2.24a)), unless the parameter c is rational with $q_c = k_c - 1$ so that $1/(1-c) = k_c$ is a *positive integer* (see (2.27b)): in which case—as in the case $c = 1$—*all* the solutions $\gamma(\tau)$ of the ODE (2.20) are *entire* functions of the *complex* variable τ and *all* the solutions $y(t)$ of the ODE (2.22) are *nonsingular* and *periodic* with, at most (see Remark 2.4.1.2), period kT, see (2.19b), (2.18b) and (2.16b).

A case with $c \neq 1$ worth of special notice is that with

$$c = 1 + \frac{1}{2\,r} = \frac{1+2\,r}{2\,r} \tag{2.30a}$$

when the ODE (2.22) reads

$$\ddot{y} = \frac{r}{2}\,\omega^2\,y + \left(\frac{1+2\,r}{2\,r}\right)\frac{(\dot{y})^2}{y}\,. \tag{2.30b}$$

Note the *absence* from this ODE of the *imaginary* unit $\mathbf{i}$. In this case the solution of the initial-value problem in the *real* case—with both $y(0)$ and $\dot{y}(0)$ *real*—reads as follows:

$$y(t) = y(0)\,\left\{\frac{\sin\left[\omega\,(\theta - t)\,/\,2\right]}{\sin(\omega\,\theta\,/\,2)}\right\}^{-2r},$$

$$\tan(\omega\,\theta) = \frac{2\,r\,\omega\,\dot{y}(0)\,y(0)}{\left[\dot{y}(0)\right]^2 + \left[r\,\omega\,y(0)\right]^2}\,; \tag{2.31}$$

it is therefore *nonsingular* iff $-2r$ is a *positive integer*; and then *periodic* with period $T = 2\pi\,/\,|\omega|$ if $-2r$ is an *even integer*, with period $T = 4\pi\,/\,|\omega|$ if $-2r$ is an *odd integer*.

We end Subsection 2.4.1 with a terse discussion of the change in the behavior of the solution (2.24) of the ODE (2.22) if the parameter ω, instead of being *real*, is *complex*, featuring, say, a *positive imaginary* part, $\mathrm{Im}\,[\omega] > 0$. Then, if $r > 0$, in the *remote future* $y(t)$ vanishes,

$$\lim_{t \to +\infty}\,\left[y(t)\right] = 0\,, \tag{2.32a}$$

while if $r = 0$ it tends to a *generally finite* value,

$$\lim_{t \to +\infty}\,\left[y(t)\right] = \tilde{y}(0)\,, \tag{2.32b}$$

where the value of $\tilde{y}(0)$ can be immediately read from (2.24). In the *remote past*—when $\eta(t)$ diverges exponentially—the behavior of the solution depends on the values of both parameters, r and c, and also on the initial data in the $c = 1$ case. Indeed, clearly in the $c = 1$ case the solution $y(t)$, see (2.24b), *diverges* respectively *vanishes* in the *remote past* if the *real part* of the (generally *complex*) number $\frac{\dot{y}(0)}{\mathbf{i}\,\omega\,y(0),} - r$ is *positive* respectively *negative*; while in the $c \neq 1$ case, the solution $y(t)$, see (2.24a), *diverges* respectively *vanishes* in the *remote past* if the (generally *real*) number $r + 1/(1 - c)$ is *positive* respectively *negative*.

Clearly these behaviors of the solution $y(t)$ in the *remote future* and the *remote past* are exchanged if the *imaginary part* of ω is *negative* rather than *positive*.

2.4.2 The ODE $\gamma'' = (\bar{y})^{-1} (\gamma')^2$ and Its Isochronous Variant

In Subsection 2.4.2 the solution of the ODE

$$\gamma'' = (\bar{y})^{-1} \left(\gamma'\right)^2 , \tag{2.33}$$

is reported, with $\gamma \equiv \gamma(\tau)$, primes denoting differentiations with respect to the independent variable τ, and with $\bar{y}$ an *a priori arbitrary complex* parameter ($\bar{y} \neq 0$); and the modified variant (see (2.19a), in this case with $r = 0$) of this equation,

$$\ddot{y} = \mathbf{i}\,\omega\,\dot{y} + (\bar{y})^{-1} (\dot{y})^2 , \tag{2.34}$$

featuring the *real* independent variable t ("time"), the *complex* dependent variable $y \equiv y(t)$ (and where superimposed dots denote time-differentiations) is then discussed, with particular attention to the *periodicity* properties of its solutions. Note that the ODE (2.33) may be obtained from (2.17) by firstly setting in it $\alpha = 2 + r$ and then letting $r \to 0$.

Remark 2.4.2.1. The equation of motion (2.33) is yielded by the *Hamiltonian*

$$h(p;\,\gamma) = a\,p^{\beta}\,\exp\left(\frac{\beta\,\gamma}{\bar{y}}\right) , \tag{2.35}$$

where, in the Hamiltonian context, the variable $p \equiv p(\tau)$ plays the role of *canonical momentum* conjugated to the *canonical variable* $\gamma \equiv \gamma(\tau)$, and the (*nonvanishing*) parameters a and β may be *arbitrarily assigned* ($a \neq 0$, $\beta \neq 0$, $\beta \neq 1$). ∎

It is easily seen that the solution of the initial-value problem of the ODE (2.33) reads as follows:

$$\gamma(\tau) = \gamma(0) - \bar{y}\,\ln\left(1 - \frac{\tau}{\bar{\tau}}\right) , \quad \bar{\tau} = \frac{\bar{y}}{\gamma'(0)} , \tag{2.36}$$

and the solution of the initial-value problem of the ODE (2.34) reads as follows:

$$y(t) = y(0) - \bar{y} \ln\left[\frac{\bar{\eta} - \exp(\mathbf{i}\,\omega\,t)}{\bar{\eta} - 1}\right], \quad \bar{\eta} = 1 + \frac{\mathbf{i}\,\omega\,\bar{y}}{y(0)}. \tag{2.37}$$

Clearly *all* these solutions (2.37) of the ODE (2.34)—characterized by *any initial data* $y(0)$, $\dot{y}(0)$—are (as functions of the *real* variable t) *nonsingular*; moreover, if the initial datum $\dot{y}(0)$ implies the *inequality* $|\bar{\eta}| > 1$, see (2.37), this solution is *periodic* with the basic period T (see (2.18b)), $y(t + T) = y(t)$; while if the initial datum $\dot{y}(0)$ implies the *complementary inequality* $|\bar{\eta}| < 1$, this solution (2.37) is *not periodic*, featuring instead the property $y(t + T) = y(t) - 2\mathbf{i}\pi\,\bar{y}$ (hence in this case this solution *diverge almost linearly* in the *remote* past and future); and note, finally, that the *nongeneric* initial datum $\dot{y}(0) = 0$ yields the trivial solution $y(t) - y(0)$.

We end Subsection 2.4.2 with a terse discussion of the change in the behavior of the solution (2.24) if the parameter ω, instead of being *real*, is *complex*, featuring, say, a *positive imaginary* part, $\text{Im}[\omega] > 0$. Then, in the *remote future* (i.e., as $t \to +\infty$)

$$y(t) = y(0) - \bar{y} \ln\left[1 + \frac{\dot{y}(0)}{\mathbf{i}\,\omega\,\bar{y}}\right] + O\left[\exp(-|\text{Im}[\omega]\,t|)\right], \tag{2.38a}$$

while in the *remote past* (i.e., as $t \to -\infty$)

$$y(t) = -\mathbf{i}\,\omega\,\bar{y}\,t + y(0) + \bar{y} \ln\left(-\frac{\mathbf{i}\,\omega\,\bar{y}}{\dot{y}(0)}\right) + O\left[\exp(-|\text{Im}[\omega]\,t|)\right]. \tag{2.38b}$$

These behaviors in the *remote future* and *past* are exchanged if instead $\text{Im}[\omega] < 0$.

2.4.3 *The ODE* $\gamma'' = c\left(\gamma'\right)^{\alpha}$ *and Its Isochronous Variant*

In Subsection 2.4.3, the solution of the ODE

$$\gamma'' = c\left(\gamma'\right)^{\alpha} \tag{2.39}$$

is reported, with $\gamma \equiv \gamma(\tau)$, primes denoting differentiations with respect to the independent variable τ, c an arbitrary (possibly *complex*) parameter, and α an *arbitrary real rational* parameter ($\alpha \neq 2$, since the case with $\alpha = 2$ has already been treated in the preceding Subsection 2.4.2). Note that this ODE corresponds to the special case of (2.16) with

$$r = \frac{\alpha - 2}{1 - \alpha}, \quad \alpha = \frac{r + 2}{r + 1} \tag{2.40}$$

in (2.17); hence, hereafter we also exclude from consideration the values $\alpha = 1$ and $r = -1$ (besides the value $\alpha = 2$, treated in the preceding Subsection 2.4.2); and we use hereafter—within Subsection 2.4.3—the parameter r (with the restriction

$r \neq 0$, $r \neq -1$) instead of the parameter α, on the understanding that they are always related by these formulas (2.40). The corresponding version of the modified equation satisfied by the time-dependent function $y(t)$ related to $\gamma(\tau)$ by (2.18) reads as follows:

$$\ddot{y} = \mathbf{i}\,(2\,r+1)\;\omega\,\dot{y} + r\,(r+1)\;\omega^2\,y + c\,\left[\dot{y} - \mathbf{i}\,r\,\omega\,y\right]^{(r+2)/(r+1)} . \qquad (2.41)$$

It is easily seen that the solution of the initial-value problem of the ODE (2.39) reads as follows:

$$\gamma(\tau) = \gamma(0) + \frac{\bar{\tau}\,\gamma'(0)}{r}\left[\left(1 - \frac{\tau}{\bar{\tau}}\right)^{-r} - 1\right] , \quad \bar{\tau} = \left(\frac{1+r}{c}\right)\,\left[\gamma'(0)\right]^{-1/(1+r)} , \qquad (2.42)$$

and the solution of the initial-value problem of the ODE (2.41) reads as follows:

$$y(t) = \tilde{y}(\eta) , \quad \eta \equiv \eta(t) = \exp(\mathbf{i}\,\omega\,t) , \qquad (2.43\text{a})$$

$$\tilde{y}(\eta) = \eta^r \left\{ y(0) + \left(\frac{1+r}{c\,r}\right)\,\left[\dot{y}(0) - \mathbf{i}\,r\,\omega\,y(0)\right]^{r/(1+r)} \cdot \right.$$
$$\left. \cdot \left[\left(\frac{\bar{\eta} - \eta}{\bar{\eta} - 1}\right)^{-r} - 1\right]\right\} , \qquad (2.43\text{b})$$

$$\bar{\eta} = 1 + \mathbf{i}\,\omega\,\left(\frac{1+r}{c}\right)\,\left[\dot{y}(0) - \mathbf{i}\,r\,\omega\,y(0)\right]^{-1/(1+r)} . \qquad (2.43\text{c})$$

Hence these solutions (2.43)—as a function of the time t—are *nonsingular* for *all* initial data $\dot{y}(0)$, $y(0)$ if r is a *negative integer*, and also for *arbitrary real* r ($r \neq 0$, $r \neq -1$) except for the *nongeneric* set of initial data implying $|\bar{\eta}| = 1$, in which case (unless r is a *negative integer* other than -1) they hit a singularity at $t = t_s$, with $t_s = (\mathbf{i}\,\omega)^{-1}\,\ln(\bar{\eta})$; they are otherwise *periodic*, with at most (see Remark 2.4.1.2) period kT, provided r is a *real rational* number (see (2.16b), (2.18b) and (2.19b)).

We end this Subsection 2.4.3 with a terse discussion of the change in the behavior of the solution (2.43) if the parameter ω, instead of being *real*, features, say, a *positive imaginary* part, Im $[\omega] > 0$. Then, depending on the sign of r, in the remote future or past $y(t)$ *vanishes*,

$$\lim_{t \to +r\,\infty}\left[y(t)\right] = 0 , \qquad (2.44\text{a})$$

or tends to a *finite* value,

$$\lim_{t \to -r\,\infty}\left[y(t)\right] = \tilde{y}(0) , \qquad (2.44\text{b})$$

where the value of $\tilde{y}(0)$ can be easily read from (2.43b) with (2.43c).

2.4.4 The Solvable ODE $\gamma'' = \rho \left(\gamma'\right)^2 \gamma^{-1} + c\gamma^\rho$ and Related Isochronous Versions

In Subsection 2.4.4 the solution of the ODE

$$\gamma'' = \rho \left(\gamma'\right)^2 \gamma^{-1} + c\,\gamma^\rho , \tag{2.45}$$

is reported. Here $\gamma \equiv \gamma(\tau)$, primes denote differentiations with respect to the independent variable τ, c is an *arbitrary* (*nonvanishing*, possibly *complex*) parameter, and ρ an *arbitrary real rational* parameter. Note that this ODE features in its right-hand side the sum of two terms (with the *same* parameter ρ playing a *different* role In these two terms). Also note that both terms satisfy the scaling property (2.16c), see (2.17): the first with $\alpha = 2$ and r *unrestricted*, the second with $\alpha = 0$ and

$$r = \frac{2}{\rho - 1} , \quad \rho = 1 + \frac{2}{r} , \tag{2.46}$$

a relationship that is maintained throughout this Subsection 2.4.4 (implying the assumption $\rho \neq 1$).

Remark 2.4.4.1. The ODE (2.45) is yielded by the *Hamiltonian*

$$h(p;\,\gamma) = a\,p^2\,\gamma^{2\rho} - \frac{c\,\gamma^{1-\rho}}{2\,a\,(1-\rho)} , \tag{2.47}$$

where, in the Hamiltonian context, the variable $p \equiv p(\tau)$ plays the role of *canonical momentum* conjugated to the *canonical variable* $\gamma \equiv \gamma(\tau)$, and the (*nonvanishing*) parameter a may be *arbitrarily assigned* ($a \neq 0$, $\rho \neq 0$, $\rho \neq 1$). ∎

The diligent reader will easily verify that the solution of the initial value problem for this ODE, (2.45), reads as follows: if $\rho \neq 1$,

$$\gamma(\tau) = \gamma(0) \left[1 + (1-\rho)\left\{\frac{\gamma'(0)\,\tau}{\gamma(0)} + \frac{c}{2}\left[\gamma(0)\right]^{\rho-1}\,\tau^2\right\}\right]^{1/(1-\rho)} ; \tag{2.48a}$$

if $\rho = 1$,

$$\gamma(\tau) = \gamma(0)\,\exp\left[\frac{\gamma'(0)\,\tau}{\gamma(0)} + \frac{c}{2}\,\tau^2\right] . \tag{2.48b}$$

But, as already noted, $\rho = 1$ implies $r = \infty$, see (2.46); so—as already indicated above—we restrict our consideration to the case with $\rho \neq 1$.

The corresponding version of the modified equation satisfied by the time-dependent function $y(t)$ related to $\gamma(\tau)$ by (2.18) reads as follows:

$$\ddot{y} = -3\,\mathbf{i}\,\omega\,\dot{y} - r\,\omega^2\,y + \left(\frac{r+2}{r}\right)\frac{(\dot{y})^2}{y} + c\,y^{(r+2)/r} , \tag{2.49}$$

and the solution of the corresponding initial-value problem reads as follows:

$$y(t) = \tilde{y}(\eta) , \quad \eta \equiv \eta(t) = \exp(\mathbf{i}\,\omega\,t) , \tag{2.50a}$$

$$\tilde{y}(\eta) = y(0) \left[b \left(1 - \frac{\eta_+}{\eta} \right) \left(1 - \frac{\eta_-}{\eta} \right) \right]^{-r/2} , \tag{2.50b}$$

$$\eta_\pm = \frac{-2 + a + 2\,b \pm \Delta}{2\,b} , \quad \Delta^2 = (a-2)^2 - 4\,b , \tag{2.50c}$$

$$a = \frac{2\,\dot{y}(0)}{\mathbf{i}\,r\,\omega\,y(0)} , \quad b = \frac{c\,[y(0)]^{2/r}}{r\,\omega^2} . \tag{2.50d}$$

As for the time evolution of this solution, there are clearly various possibilities depending on the parameters and on the initial data.

Let us consider firstly the *nongeneric* set of initial data such that either $|\eta_+| = 1$ or $|\eta_-| = 1$ or $|\eta_+| = |\eta_-| = 1$, see (2.50c). Then in its time evolution the solution $y(t)$ shall hit a *singularity*—at the time t_S such that $\eta(t_S) = \eta_+$ or $\eta(t_S) = \eta_-$, as the case may be; unless the parameter r is a *negative even integer* (or *any negative integer* if moreover $\eta_+ = \eta_-$, i.e., $\Delta = 0$; see (2.50c)), in which case clearly—for *any* initial data—the solution $y(t)$ is *nonsingular* for *all* time and *periodic* with at most (see Remark 2.4.1.2) period kT (see (2.18b), (2.16b) and (2.19b)).

Let us then turn our attention to the *generic* case of initial data such that $|\eta_+| \neq 1$ and $|\eta_-| \neq 1$, and moreover r is *not* an *even integer* and $\eta_+ \neq \eta_-$. There are then *three* possibilities, depending on the values of the initial data $y(0)$ and $\dot{y}(0)$.

Case (i). If the initial data $y(0)$ and $\dot{y}(0)$ imply

$$|\eta_+| > 1 , \quad |\eta_-| > 1 , \tag{2.51a}$$

the two branch points of $\tilde{y}(\eta)$—in the *complex* η-plane, at $\eta = \eta_+$ and $\eta = \eta_-$, see (2.50b)—fall *outside* the circle $\tilde{C}$, centered at the origin and of *unit* radius, on which travels the point $\eta(t) = \exp(\mathbf{i}\,\omega\,t)$ as a function of the time t. Hence in this case the solution $y(t)$ is *periodic* with period kT (see (2.18b) and (2.16b)), due to the *rational* branch point (of order k) at the center, $\eta = 0$, of the circle $\tilde{C}$.

Case (ii). If the initial data $y(0)$ and $\dot{y}(0)$ imply

$$|\eta_+| > 1 , \quad |\eta_-| < 1 , \quad \text{or} \quad |\eta_+| < 1 , \quad |\eta_-| > 1 , \tag{2.51b}$$

then the right-hand side of (2.50b) features, inside the circle $\tilde{C}$—besides the branch point at $\eta = 0$—another *rational* branch point (of order $1/(-2k)$), hence the solution $y(t)$ is *periodic* with period $2kT$ (see (2.18b) and (2.16b)).

Case (iii). If the initial data $y(0)$ and $\dot{y}(0)$ imply

$$|\eta_+| < 1 , \quad |\eta_-| < 1 , \tag{2.51c}$$

then the right-hand side of (2.50b) features, inside the circle $\tilde{C}$—besides the branch point at $\eta = 0$—two branch points at $\eta = \eta_+$ and $\eta = \eta_-$, and since due to each of these two branch points after each round around the circle $\tilde{C}$—which takes a time T, see (2.18b) and the definition $\eta(t) = \exp(\mathbf{i}\,\omega\,t)$, see (2.50a)—the solution $y(t)$ gets multiplied by a factor $\exp(\mathbf{i}r\omega t/2)$ while due to the branch point at $\eta = 0$ the solution $y(t)$ gets multiplied by a factor $\exp(-\mathbf{i}r\omega t)$ in this case the solution $y(t)$ turns out to be periodic with the basic period T, see (2.18b).

Note that this implies that, for *arbitrary nongeneric* initial data—such that $|\eta_+| \neq 1$ and $|\eta_-| \neq 1$—the solution $y(t)$ of the ODE (2.49), with an *arbitrary rational* assignment of the parameter $r = q/k$ (see (2.16b)), is *periodic* with at most (see Remark 2.4.1.2) period kT; this ODE is therefore *isochronous* (provided the parameter ω is an *arbitrary nonvanishing real* number).

If, instead, the parameter ω is *not real* but does feature, say, a *positive imaginary* part, $\mathrm{Im}\,[\omega] > 0$, then clearly the periodicity properties of the solution $y(t)$ mentioned above do not hold any more. In this case, as $t \to +\infty$, $\eta(t) = \exp(\mathbf{i}\,\omega\,t)$ tends exponentially to zero, while as $t \to -\infty$, $\eta(t) = \exp(\mathbf{i}\,\omega\,t)$ diverges exponentially; and the corresponding *asymptotic behaviors* as $t \to \pm\infty$ of the solution $y(t)$ can be read immediately from its expression (2.50).

2.4.5 Other Solvable Cases of the ODE $\gamma'' = c\,(\gamma')^\alpha\,\gamma^\beta$ and Related Isochronous Variants

In Subsection 2.4.5 we consider the ODE

$$\gamma'' = c\,(\gamma')^\alpha\,\gamma^\beta \tag{2.52a}$$

(where, as above, $\gamma \equiv \gamma(\tau)$ and appended primes indicate τ-derivatives), mainly in the special case with

$$\alpha = \frac{3+\beta}{2+\beta}\,, \quad \beta = \frac{2\alpha-3}{1-\alpha}\,. \tag{2.52b}$$

In this case this ODE is *solvable* in terms of elementary functions.

Remark 2.4.5.1. In Subsection 2.4.5 we exclude from consideration the case with $\alpha = 2$, $\beta = -1$, which has already been treated in Subsection 2.4.1; while the case with $\alpha = 0$, $\beta = -3$ is a special case of the model discussed below in Subsection 4.2.4. These are the only cases in which both α and β are *integers* (consistently with (2.52b)). For obvious reasons (see (2.52b)), in the first part of this Subsection 2.4.5 we assume $\alpha \neq 1$ and $\beta \neq -2$; but at the end of this Subsection 2.4.5, we also identify some *solvable* ODEs of type (2.52a) with $\alpha = 1$ but without (2.52b). ∎

The diligent reader will have no difficulty to verify that the solution of the initial value problem for this ODE, (2.52), reads as follows:

$$\gamma(\tau) = a^r \left\{ a^{-1/r} (\tau + b)^{-1/r} - C \right\}^{-r}, \tag{2.53a}$$

$$a = \left[\gamma'(0) \right]^{1/(1+r)} - C \left[\gamma(0) \right]^{1/r}, \tag{2.53b}$$

$$b = a^{-1} \left\{ a \left[\gamma(0) \right]^{-1/r} + C \right\}^{-r}, \tag{2.53c}$$

$$C = \frac{c\,r}{1+r}, \quad r = \frac{1-\alpha}{\alpha-2} = \frac{1}{1+\beta}. \tag{2.54}$$

The (autonomous!) *isochronous* variant of this ODE obtains via the usual position (see above, at the beginning of this Section 2.4),

$$y(t) = \exp(\mathbf{i}\,r\,\omega\,t)\,\gamma(\tau), \quad \tau \equiv \tau(t) = \frac{\exp(\mathbf{i}\,\omega\,t) - 1}{\mathbf{i}\,\omega} \tag{2.55}$$

with r defined as above, see (2.55), and it reads as follows:

$$\ddot{y} = \mathbf{i}\,(2\,r+1)\,\omega\,\dot{y} + r\,(r+1)\,\omega^2\,y$$
$$+ c\,(\dot{y} - \mathbf{i}\,r\,\omega\,y)^{(1+2r)/(1+r)}\,y^{(1-r)/r}. \tag{2.56}$$

The solution of this ODE reads as follows:

$$y(t) = \tilde{y}(\eta), \quad \eta \equiv \eta(t) = \exp(\mathbf{i}\,\omega\,t), \tag{2.57a}$$

$$\tilde{y}(\eta) = \left(\frac{a\,\eta}{C} \right)^r \left[\left(\frac{\eta_1 - \eta_2}{\eta_1 - \eta} \right)^{1/r} - 1 \right]^{-r}, \tag{2.57b}$$

$$\eta_1 = 1 - \mathbf{i}\,\omega\,b, \quad \eta_2 = \eta_1 + \mathbf{i}\,\omega\,a^{-1}\,C^{-r}. \tag{2.57c}$$

Let us tersely discuss the time evolution of this solution. Note that the function $\tilde{y}(\eta)$ generally features 4 *rational* branch points, at $\eta = 0$, $\eta = \eta_1$, $\eta = \eta_2$, $\eta = \infty$, while as a function of the time t the *complex* variable $\eta(t)$ travels on the *circle* $\tilde{C}$—centered at the origin of the *complex* η-plane, and having *unit* radius—taking the time T, see (2.18b), to perform a full round on this circle. Hence, if the two branch points at $\eta = \eta_1$ and $\eta = \eta_2$ fall *outside* this circle $\tilde{C}$—i.e., if the initial data $y(0)$ and $\dot{y}(0)$ imply

$$|\eta_1| > 1, \quad |\eta_2| > 1, \tag{2.58a}$$

see (2.57c)—then the only relevant branch point is that at the origin hence the solution $y(t)$ is periodic with period kT, see (2.18b), (2.16b) and (2.19b). And since the set of initial data corresponding to the *inequalities* (2.58a) is clearly an *open* set, this is enough to qualify the ODE (2.56) as *isochronous*.

Likewise, if the initial data imply that the two inequalities (2.58a) are reversed, i.e.,

$$|\eta_1| < 1 , \quad |\eta_2| < 1 , \qquad (2.58b)$$

then clearly the only relevant branch point is that at infinity (associated with the exponent $r+1 = (q + k) /k$, see (2.16b), hence of order k), hence again the solution $y(t)$ is periodic with period kT, see (2.18b), (2.16b) and (2.19b).

The interested reader will have no difficulty to analyze the cases when these *equalities* are replaced by *equalities* or only one of them is *reversed*.

We end this discussion of the solutions (2.57) of the ODE (2.56) with a terse mention of the drastic change in their behaviors that is obtained if the restriction that the parameter ω be *real* is abandoned and is instead replaced, say, by the assumption that it feature a *positive imaginary* part, Im $[\omega] > 0$. Then clearly, if $r > 0$, in the remote future $y(t)$ tends (exponentially) to zero, $y(\infty) = 0$, while in the remote past it diverges *exponentially*; with these behaviors exchanged if $r < 0$.

Finally, let us consider the variant of the (autonomous) ODE (2.52a) (without (2.52b)) with $\alpha = 1$ and β *a priori arbitrary* (but see below),

$$\gamma'' = c \, \gamma' \, \gamma^\beta . \qquad (2.59a)$$

Its (autonomous!) *isochronous* variant obtains via the change of dependent variables (2.55) with

$$r = 1/\beta \qquad (2.59b)$$

hence it reads as follows:

$$\ddot{y} = \mathbf{i} \, (2 \, r + 1) \, \omega \, \dot{y} + r \, (r + 1) \, \omega^2 \, y + c \, (\dot{y} - \mathbf{i} \, r \, \omega \, y) \, y^{1/r} . \qquad (2.60)$$

Remark 2.4.5.2. The ODE (2.59a) is yielded, for $\beta \neq -1$ (hence $r \neq -1$) by the Hamiltonian

$$h \, (p, \, \gamma) = f \, (\gamma) \, \exp \, (a \, p) - \frac{c \, \gamma^{\beta+1}}{a \, (\beta + 1)} , \qquad (2.61a)$$

and, for $\beta = -1$ (hence $r = -1$) by the Hamiltonian

$$h \, (p, \, \gamma) = f \, (\gamma) \, \exp \, (a \, p) + \ln \left(\gamma^{-c/a} \right) . \qquad (2.61b)$$

In these formulas, p is the canonical momentum conjugated to the canonical variable γ, while a is an *arbitrary nonvanishing* parameter and $f \, (\gamma)$ an *arbitrary differentiable* function (of course *not identically vanishing*). ■

The ODE (2.59a) can be immediately integrated once, yielding

$$\gamma'(\tau) = \left(\frac{c}{1+\beta}\right) \left[\gamma(\tau)\right]^{1+\beta} + B,$$

$$B = \gamma'(0) - \left(\frac{c}{1+\beta}\right) \left[\gamma(0)\right]^{1+\beta}, \qquad (2.62)$$

and a second integration can be performed in more or less *explicit* form in terms of elementary functions for various values of β. For simplicity, we hereafter restrict attention to the $\beta = 1$ (hence $r = 1$) case, leaving to the interested reader the treatment of other cases. Then the two ODEs (2.59a) respectively (2.60) read simply

$$\gamma'' = c\,\gamma'\,\gamma, \qquad (2.63)$$

respectively (since $\beta = 1$ implies $r = 1$)

$$\ddot{y} = 3\,\mathbf{i}\,\omega\,\dot{y} + 2\,\omega^2\,y + c\,(\dot{y} - \mathbf{i}\,\omega\,y)\,y. \qquad (2.64)$$

The complete integration of the first, (2.63), of these two ODEs is then a trivial task, yielding

$$\gamma(\tau) = \frac{\gamma(0) + a\,\tanh\left(\frac{a\,c\,\tau}{2}\right)}{1 + \frac{\gamma(0)}{a}\,\tanh\left(\frac{a\,c\,\tau}{2}\right)},$$

$$a^2 = \left[\gamma(0)\right]^2 - \frac{2\,\gamma'(0)}{c}. \qquad (2.65)$$

Note that this expression of $\gamma(\tau)$ depends on a^2 rather than a.

The solution of the initial-value problem for the second, (2.64), of these two ODEs reads

$$y(t) = \exp(\mathbf{i}\,\omega\,t)\,\gamma(\tau), \quad \tau = \frac{\exp(\mathbf{i}\,\omega\,t) - 1}{\mathbf{i}\,\omega}, \qquad (2.66a)$$

where $\gamma(\tau)$ is defined as above, see (2.65), but with

$$\gamma(0) = y(0), \quad a^2 = \left[y(0)\right]^2 - \frac{2\left[\dot{y}(0) - \mathbf{i}\,\omega\,y(0)\right]}{c}. \qquad (2.66b)$$

Hence *all* the solutions $y(t)$ of this ODE (2.64) are *periodic*, as functions of the *real* variable t, with the basic period $T = 2\pi/|\omega|$, and only a subset of these solutions feature a *polar* singularity as t evolves from $-\infty$ to $+\infty$.

2.N Notes on Chapter 2

The formulas (2.7) and (2.8) are reported and proven in [34]; the formulas (2.9) and (2.10) are reported and proven in [3]; analogous formulas for higher derivatives (of arbitrary order k; and, in explicit form, for k up to 6) are reported and proven in [12].

For the notion of *asymptotic periodicity* mentioned in Remark 2.3.2—and the related notion of *asymptotic isochrony*—see [54] and Chapter 6 of [27].

For the extension of the main treatment from (monic) *polynomials* of degree N hence featuring N *coefficients* and N *zeros* to *entire functions* featuring an *infinite* number of *coefficients* and *zeros*, respectively to appropriately normalized *rational* functions featuring N *poles* and N *residues*, see F. Calogero, "Zeros of entire functions and related systems of infinitely many nonlinearly coupled evolution equations", Theor. Math. Phys. **196** (2), 1111–1128 (2018), respectively F. Calogero, "Zeros of Rational Functions and Solvable Nonlinear Evolution Equations", J. Math. Phys. **59** (2018) (in press).

The *trick* used in Section 2.4 to generate ODEs featuring periodic solutions was introduced in [19] and has been extensively used since, see in particular [27] and references therein.

Finally, let us emphasize that the ODEs discussed in Section 2.4—*solvable* in terms of elementary functions—by no means exhaust the universe of such examples: the alert and interested reader will easily identify many others. The few examples reported and tersely discussed in Section 2.4 are mainly introduced as material to be utilized below, see in particular Chapter 4.

3

A Differential Algorithm to Compute All the Zeros of a Generic Polynomial

The investigation of the *zeros* of a given polynomial—itself defined by assigning its *coefficients*—is a foundational problem of mathematics. A fundamental result states that—in the *complex* domain—every polynomial of degree N features N *zeros* (counting, if need be, their *multiplicities*). The development and investigation of techniques to *compute* the *zeros* of a polynomial of degree $N > 4$—the cases when this task cannot be generally managed via *explicit* formulas involving elementary functions such as quadratic, cubic or quartic roots—has exercised the expertise of mathematicians for many centuries. In Chapter 3 we report a recently introduced technique—based on a *differential algorithm*—to perform this task. The first version of this technique, which involves the solution of a system of *first-order* ODEs, is described in Section 3.1. A second version of this technique—based instead on a system of *second-order* ODEs, but perhaps preferable due to its remarkable neatness—is described in Section 3.2. These two sections are drafted so as to make their *separate* reading possible; hence the reader who reads both might notice some repetitions ...

3.1 A Differential Algorithm Based on a System of First-Order ODEs Suitable to Evaluate All the Zeros of a Generic Polynomial

Let the *generic* polynomial

$$P_N\left(z;\,\vec{c},\,\tilde{z}\right) = z^N + \sum_{m=1}^{N}\left(c_m\,z^{N-m}\right) = \prod_{n=1}^{N}\left(z - z_n\right) \tag{3.67a}$$

be characterized by its N *coefficients* c_m and by its N *zeros* z_n, implying (see (2.4))

$$c_m = (-1)^m\,\sigma_m\left(\tilde{z}\right)\ . \tag{3.67b}$$

A procedure to compute these N zeros z_n is then provided by the following:

Proposition 3.1.1. Consider the following system of N *nonlinear first-order differential equations* satisfied by the N *zeros* $x_n(t)$ of a t-dependent polynomial such as (2.3):

$$\dot{x}_n(t) = -\left\{\prod_{\ell=1,\,\ell\neq n}^{N}[x_n(t)-x_\ell(t)]^{-1}\right\}$$

$$\sum_{m=1}^{N}\left\{g_m(t;T)\left[c_m-(-1)^m\,\sigma_m(\tilde{x}(0))\right][x_n(t)]^{N-m}\right\},\qquad(3.68a)$$

where (see (2.4b))

$$\sigma_m(\tilde{x}(0)) = \sum_{1\leq n_1<n_2<\cdots<n_m\leq N}\left[x_{n_1}(0)\cdot x_{n_2}(0)\cdots x_{n_m}(0)\right]$$

$$= \frac{1}{m!}\sum_{n_1,n_2,\ldots n_m=1}^{N}\left[x_{n_1}(0)\cdot x_{n_2}(0)\cdots x_{n_m}(0)\right],\qquad(3.68b)$$

and

$$g_m(t;T) = \frac{\dot{f}_m(t)}{f_m(T)-f_m(0)},\quad\text{implying}\quad\int_0^T dt\,g_m(t;T) = 1.\qquad(3.68c)$$

Here the N (possibly *complex*) functions $f_m(t)$, and the single *positive* parameter T, can be assigned essentially *arbitrarily*; but of course so that the N functions $g_m(t)$ be *finite* for $0\leq t\leq T$ hence feature the property displayed by the second equality (3.68c) (and let us recall that superimposed dots denote t-differentiation).
Then

$$z_n = x_n(T).\ \blacksquare\qquad(3.68d)$$

The proof of this result is an immediate consequence of the identity (2.7), as seen by setting in it

$$y_m(t) = (-1)^m\,\sigma_m(\tilde{x}(0))+\left[\frac{f_m(t)-f_m(0)}{f_m(T)-f_m(0)}\right]\left[c_m-(-1)^m\,\sigma_m(\tilde{x}(0))\right].\ (3.69)$$

Indeed, this assignment clearly implies that the system (2.7) coincides—via the definition (3.68c) of $g_m(t;T)$—with the system (3.68a), while the coefficients (3.69) (of the polynomial (2.3)) clearly interpolate from the *initial* values $y_m(0) = (-1)^m\,\sigma_m(\tilde{x}(0))$ at $t=0$ (consistently with (3.68b)) to the values $y_m(T) = c_m$, implying that at $t=T$ the polynomial (2.3) coincides with the polynomial (3.67a). Hence, its zeros $x_n(T)$ coincide with the zeros z_n of this polynomial (3.67a), validating the result (3.68d). Q. E. D.

It is thus seen that the N zeros z_n of the polynomial $P_N(z;\vec{c},\tilde{z})$, see (3.67a), can be computed—once the N coefficients c_m of this polynomial have been assigned—via the following procedure. *Step one*: choose (*arbitrarily!*) N *complex* numbers $x_n(0)$.

Step two: compute, via the formulas (3.68b), the N quantities $\sigma_m\,(\tilde{x}\,(0))$. *Step three*: integrate (*numerically*) the system of differential equations (3.68a) from $t = 0$ to $t = T$, starting from the N initial data $x_n\,(0)$, thereby getting the N values $x_n\,(T)$, which give the sought result, see (3.68d).

Let us emphasize the ample degrees of *freedom* of this procedure implied by the *arbitrariness* in assigning the N initial data $x_n\,(0)$, the N functions $f_m\,(t)$, and the parameter T.

Will this procedure always work? The only possible snag is that the solution $x_n\,(t)$ of the "dynamical system" (3.68a) run into a *singularity* during its evolution from $t = 0$ to $t = T$. The mechanism whereby this might occur is because during this evolution two different coordinates $x_n\,(t)$ might coincide, $x_\ell\,(t) = x_n\,(t)$ for $\ell \neq n$, at some value of the *real* variable t in the interval $0 < t < T$, causing the right-hand side of (3.68a) to blow up. This "collision" might indeed happen, but it is *not* a *generic* phenomenon: hence it will be enough to change the assignment of the (*arbitrary!*) initial data $x_n\,(0)$—or of the N functions $f_m\,(t)$ or of the parameter T—to avoid this difficulty; note however that this suggests that to apply this method it will be advisable to start with *complex* initial data $x_n\,(0)$, even in the case of *real* polynomials with *real* zeros. And note, moreover, that the numerical integration of the differential equations (3.68a) with different initial data $x_n\,(0)$ and with different assignments of the N functions $f_m\,(t)$ and of the parameter T—for instance,

$$f_m\,(t) = t\,, \quad T = 1\,, \quad \text{implying } g_m\,(t) = 1\,, \tag{3.70a}$$

or

$$f_m\,(t) = a_m\,t^{\lambda_m}\,, \quad \lambda_m > 0\,, \quad T = 1\,, \quad \text{implying } g_m\,(t) = \lambda_m\,t^{\lambda_m - 1} \tag{3.70b}$$

with N arbitrary, possibly *complex*, parameters λ_m, or

$$f_m\,(t) = \frac{\exp\,(\lambda_m\,t) - 1}{\exp\,(\lambda_m\,T) - 1}\,, \quad \lambda_m \neq 0\,, \quad \text{implying } g_m\,(t) = \frac{\lambda_m\,\exp\,(\lambda_m\,t)}{\exp\,(\lambda_m\,T) - 1}\,, \tag{3.70c}$$

again with N arbitrary, possibly *complex*, parameters λ_m, while T is an *arbitrary positive* number—allows to assess the *accuracy* of the computation, by comparing the results obtained with different assignments of these input data.

Remark 3.1.1.   This procedure will work more efficiently the closer the, *arbitrarily* chosen, *initial* values $x_n\,(0)$ are to the N *zeros* z_n the values of which one is trying to compute; indeed if all the N initial values $x_n\,(0)$ happened to *coincide* with (an appropriate permutation of) the N zeros z_n so that $x_n\,(0) = z_n$, this would imply $y_m\,(0) = c_m$ hence the right-hand side of the differential equations (3.68a) would *vanish* (see (2.4), (3.68b) and (3.67b)), entailing $\dot{x}_n = 0$ hence $x_n\,(T) = x_n\,(0) = z_n$, consistently with (3.68d). This suggests that, in the context of the *numerical* task to compute the N zeros z_n, it might be convenient to employ this technique *iteratively*, using the output $\tilde{x}(T)$ of each application of it as input $\tilde{x}(0)$ for a subsequent application. ∎

Let us also note that the dependence (via (3.68b)) of the right-hand sides of the differential equations (3.68a) upon the initial values $x_n(0)$ of the dependent variables $x_n(t)$ implies that these are rather *Differential Functional Equations* than *Ordinary Differential Equations*; but this fact has no relevance on the procedure described above to compute *all* the N zeros of a given (monic, *generic*) polynomial of *arbitrary* degree N.

3.2 A Differential Algorithm Based on a System of Second-Order ODEs Suitable to Evaluate All the Zeros of a Generic Polynomial

3.2.1 Introduction

The differential algorithm to evaluate *all* the zeros of a *generic* polynomial described in Section 3.1 involves the solution of an appropriate system of N nonlinearly coupled *first-order* ODEs. The novel technique described below is analogous, but it instead involves the solution of an appropriate system of N nonlinearly coupled *second-order* ODEs. While this fact might be considered a drawback of the novel system when compared to that described above, the novel technique described below is remarkably neater and therefore also worthy of being widely known at least because of its elegance.

3.2.2 The Task

Let a *generic* (monic) polynomial be defined as follows in terms of its N coefficients c_m and its N zeros x_n:

$$p_N(z) = z^N + \sum_{m=1}^{N} \left(c_m \, z^{N-m} \right) = \prod_{n=1}^{N} (z - x_n) \ . \tag{3.71}$$

Hereafter, we assume that N is an *arbitrary positive* integer N, and that the N coefficients c_m are N, *arbitrarily* assigned, generally *complex*, numbers. The task is to compute the N zeros x_n (all different among themselves: "genericity" of the polynomial (3.71)). As it is well known, this task can be generally performed *explicitly* only for $N = 2$, $N = 3$ respectively $N = 4$, via equations involving square, cubic, respectively quartic roots. The N coefficients c_m can instead be expressed in an *explicit* manner in terms of the N zeros x_n: $c_m = (-1)^n \, \sigma_m(\tilde{x})$, where $\sigma_m(\tilde{x})$ is the symmetrical sum of order m of the N zeros x_n which are the N elements of the *unordered* set $\tilde{x}$; but, remarkably, these formulas play no role in the following.

Hereafter indices such as n, m, ℓ run over the *positive integers* from 1 to N, unless otherwise indicated.

3.2.3 The Algorithm

Consider the dynamical system characterized by the following N *second-order* ODEs (of Newtonian type: "accelerations equal forces")

$$\ddot{\xi}_n(t) = \sum_{\ell=1;\ \ell\neq n}^{N} \left[\frac{2\, \dot{\xi}_n(t)\, \dot{\xi}_\ell(t)}{\xi_n(t) - \xi_\ell(t)} \right], \qquad (3.72)$$

which characterizes the motion in the complex ξ-plane of the N points identified by the N coordinates $\xi_n(t)$; here, and below, superimposed dots indicate of course differentiation with respect to the *real* parameter t ("dimensionless time": the independent variable). This system (3.72) is, in fact, an *integrable Hamiltonian* system, to which the honorary name "goldfish" has been attributed because of the neatness of its equations of motion, which are clearly autonomous and invariant under any arbitrary common translation or rescaling of the dependent variables $\xi_n(t)$ (see Section 4.1).

Next, integrate this system (3.72) of N second-order ODEs from $t=0$ to $t=1$, starting—at the "initial time" $t=0$—from N, *arbitrarily* assigned, generally *complex*, initial values $\xi_n(0)$, with the corresponding initial velocities $\dot{\xi}_n(0)$ given by the following *explicit* formula in terms of these N initial data $\xi_\ell(0)$ and of the N coefficients c_m of the polynomial (3.71):

$$\dot{\xi}_n(0) = -\frac{[\xi_n(0)]^N + \sum_{m=1}^{N}\left\{ c_m\, [\xi_n(0)]^{N-m} \right\}}{\prod_{\ell=1,\ \ell\neq n}^{N}[\xi_n(0) - \xi_\ell(0)]} = -\frac{p_N[\xi_n(0)]}{\prod_{\ell=1,\ \ell\neq n}^{N}[\xi_n(0) - \xi_\ell(0)]}.$$
$$(3.73)$$

There thus obtain the N zeros x_n of the polynomial (3.71):

$$\xi_n(1) = x_n, \quad p_N(x_n) = 0. \qquad (3.74)$$

3.2.4 The Proof

Let the time-dependent monic polynomial

$$P(z;t) = z^N + \sum_{m=1}^{N}\left[\gamma_m(t)\, z^{N-m} \right] = \prod_{n=1}^{N} [z - \xi_n(t)] \qquad (3.75)$$

be characterized by its N time-dependent coefficients $\gamma_m(t)$ and by its N zeros $\xi_n(t)$. In Subsection 2.2.1 it is shown that the first and second time-derivatives of these quantities are related by the following convenient *identities* (for notational simplicity we omit below to indicate *explicitly* the *t*-dependence of the various quantities, whenever this omission is unlikely to cause any misunderstanding):

$$\dot{\xi}_n = -\frac{\sum_m^N \left[\dot{\gamma}_m \, (\xi_n)^{N-m} \right]}{\prod_{\ell=1, \, \ell \neq n}^N (\xi_n - \xi_\ell)} \,, \tag{3.76a}$$

$$\ddot{\xi}_n = \sum_{\ell=1; \, \ell \neq n}^N \left(\frac{2 \, \dot{\xi}_n \, \dot{\xi}_\ell}{\xi_n - \xi_\ell} \right) - \frac{\sum_m^N \left[\ddot{\gamma}_m \, (\xi_n)^{N-m} \right]}{\prod_{\ell=1, \, \ell \neq n}^N (\xi_n - \xi_\ell)} \,; \tag{3.76b}$$

while it is plain, see (3.75), that there hold the *N identities*

$$(\xi_n)^N + \sum_{m=1}^N \left[\gamma_m \, (\xi_n)^{N-m} \right] = 0 \,. \tag{3.77}$$

Now make the assignment

$$\gamma_m \, (t) = \gamma_m \, (0) + t \left[c_m - \gamma_m \, (0) \right] \,, \tag{3.78a}$$

which is obviously valid at $t = 0$ and it clearly implies

$$\ddot{\gamma}_m \, (t) = 0 \,, \quad \dot{\gamma}_m \, (t) = c_m - \gamma_m \, (0) \,, \quad \gamma_m \, (1) = c_m \,. \tag{3.78b}$$

Clearly the first of these 3 *identities* (3.78b) implies (3.72) via (3.76b); at $t = 0$, the second of these 3 *identities* (3.78b), via (3.76a) together with (3.77), implies (3.72); and the third of these 3 *identities* (3.78b)—together with the definitions of the polynomials $p_N \, (z)$ and $P \, (z; t)$ in terms of their coefficients, see (3.71) and (3.75)) implies $p_N \, (z) = P_N \, (z; 1)$ hence the validity of (3.74). Q. E. D.

3.2.5 Remarks

Remark 3.2.5.1.   It is plain that the solutions $\xi_n \, (t)$ of the system (3.72) with arbitrary initial data $\xi_n \, (0)$ and with the initial velocities (3.73) are finite for all (finite) time—hence in particular for $0 \leq t \leq 1$—since they are the zeros of the polynomial $P_N \, (z; t)$ (see (3.75)), with the coefficients $\gamma_m \, (t)$ given by the simple prescription (3.78a). The system of nonlinear ODEs (3.72) might, however, feature a singularity— due to vanishing of the denominator in the right-hand side of (3.72)—for $0 \leq t \leq 1$ if in that time interval two (or more) of the points $\xi_n \, (t)$ moving in the complex ξ-plane collide. But clearly the set of N initial data $\xi_n \, (0)$ causing such collisions has a *vanishing* measure in the space of all sets of initial data as long as one is working with *complex* numbers (as above recommended); and any eventual collision can be avoided by a slight change of the—*a priori arbitrary*—assignment of the initial data

$\xi_n(0)$. Moreover, any eventual collision does not entail a blow-up of the solutions of (3.72)—see above—but merely, after the collision, a "loss of identity" of the colliding points; not very relevant for the evaluation of the zeros x_n of the polynomial (3.71), which have themselves no specific identity as the N elements of the *unordered* set $\tilde{x}$, see above. On the other hand a collision of two or more of the moving points $\xi_n(t)$ at $t = 1$ would be unavoidable if the polynomial $p_N(z)$ (see (3.71)) features one or more *multiple* zeros. For this reason the treatment in this book is restricted to *generic* polynomials which by definition do *not* feature *multiple* zeros. A modification of the algorithm introduced here—and of the previous analogous algorithm (see [7])—to make them suitable to deal with polynomials featuring multiple zeros is possible but it requires a separate treatment (for appropriate hints see [7]). ■

Remark 3.2.5.2.   Let us re-emphasize the flexibility of the algorithm described above entailed by the *arbitrariness* in the assignment of the N initial data $\xi_n(0)$. It is moreover plain that if any one of these N *a priori arbitrary* initial data $\xi_n(0)$ were to coincide with one of the zeros x_n of the polynomial $p_N(z)$ (see (3.71))—say, $\xi_1(0) = x_1$—then $\dot{\xi}_1(0) = 0$ would follow (see (3.72)); hence as well $\dot{\xi}_1(t) = 0$ (see (3.72)), implying that the coordinate $\xi_1(t)$ would remain fixed throughout its time evolution from $t = 0$ to $t = 1$; and moreover its presence would have no effect at all on the movement of the other points $\xi_n(t)$ with $n \neq 1$ (as clearly implied by (3.72)). ■

Remark 3.2.5.3.   The preceding Remark 3.2.5.2 and the general character of the algorithm described above suggest that the numerical efficiency of that algorithm is likely to be improved the closer the assignment of the initial data $\xi_n(0)$ are to the (*a priori* unknown) zeros x_n. This suggests the possible advantage of an iterated application of the algorithm, with the outcome of a computational cycle used as input data for the next computational cycle—possibly with an appropriate sequential increase of the precision of the numerical routine employed to integrate from $t = 0$ to $t = 1$ the system of N ODEs (3.72). ■

Remark 3.2.5.4.   While the preceding Remark 3.2.5.3 provides some potentially useful hints relevant to the applicability of the algorithm described above to actually *compute* the N zeros of a given polynomial, clearly a detailed comparison in the context of *numerical analysis* of the merits/demerits of this technique—when compared to other standard *numerical* techniques to compute all the zeros of a given polynomial—is a nontrivial endeavor which exceeds the scope of this treatment.

3.N Notes on Chapter 3

For the findings reported in Section 3.1 see [35] and the last paragraph of [36]; those reported in Section 3.2 are based largely *verbatim* on [44].

Any *new* method to evaluate the zeros of a given polynomial is likely to be fruitful independently from its numerical proficiency, as a tool to obtain interesting information over other properties of the zeros of a polynomial than their numerical values. But are the techniques reported in Chapter 3 indeed *new*?

A *cogent* answer to this question is hardly possible, since it would require a detailed scrutiny of the world mathematical literature spanning at least 4 centuries. It seems however reasonable to conjecture that these techniques are indeed *new*, for the following 3 reasons. (i) A search via the web and by word of mouth yielded no contrary indication when the technique described in Section 3.1 was recently introduced [35], [36]. (ii) This state of affairs has persisted after the publication of that technique, which was indeed advertised to be *new* [35], [36], [44]. (iii) The neatness and elegance of the algorithm presented in Section 3.2 suggests that, had the technique described there been previously discovered, it is quite unlikely that it might have been subsequently forgotten.

In this book no attempt is reported to compare the *effectiveness* of the techniques described in Chapter 3 with that of other methods to compute *all* the N zeros of a *generic* polynomial of *arbitrary* degree N. Indeed, "this is a task to be rather pursued by specialists in numerical analysis if they consider it worthy of their attention" [35]. Let us however confirm that the techniques reported in this Chapter 3 do actually work when they are numerically implemented: it is a pleasure to thank my colleagues François Leyvraz and Matteo Sommacal for having separately performed appropriate numerical checks confirming this fact and for several enlightening conversations; and to report that such numerical checks were also performed by a clearly competent, but somewhat skeptical, Referee before approving the publication of [44].

The approach to compute the zeros of a polynomial described in Chapter 3 may of course be also applied to polynomials whose zeros are *a priori* known, and in particular have a *Diophantine* character (for instance, are *integer* numbers). For such developments see F. Calogero, "Finite and infinite systems of nonlinearly coupled ordinary differential equations the solutions of which feature remarkable Diophantine findings", J. Nonlinear Math. Phys. **25**, 434–442 (2018).

I would like to thank Professor Peter Olver for pointing out to me (at the SPT2018 Workshop, June 2018) that the differential algorithms to evaluate all the zeros of a generic polynomial described in this chapter may be considered special cases of the more general and more sophisticated "homotopy continuation technique" introduced over a decade ago to investigate the zeros of systems of polynomials: See, for instance, A. J. Sommese and C. W. Wampler, *Numerical Solution of Systems of Polynomials Arising in Engineering and Science*, World Scientific, Singapore, 2005, and the literature referred to there.

4

Solvable and Integrable Nonlinear Dynamical Systems: Mainly Newtonian N-Body Problems in the Plane

In Chapter 4 we indicate how to identify many *solvable/integrable* dynamical systems. We will mainly focus on dynamical systems characterized by systems of *N second-order* Ordinary Differential Equations (ODEs) determining the motion of N point-particles moving in the *complex* plane as function of time: corresponding to many-body models of Newtonian type ("accelerations equal forces"). The main tool to uncover these models and to investigate their behavior is to identify the coordinates of the N moving particles with the N *zeros* $x_n(t)$ of polynomials such as (2.3), with the corresponding *coefficients* $y_m(t)$ themselves evolving in *solvable/integrable* manners; and then to use the key formulas of Chapter 2 (mainly those of Subsection 2.2.1) to get the corresponding evolution equations for the *zeros* $x_n(t)$. In order to illustrate the potentialities of this approach we discuss in Chapter 4—by far the longest in this book—several models, largely relying on the *solvable* ODEs discussed in Section 2.4. Hence throughout this Chapter 4 the *real* parameter t has the significance of "time", while the coordinates $x_n(t)$ are generally *complex* numbers; of course the motion in the *complex* x-plane of the points characterized by these coordinates $x_n(t)$ can be mapped into motions taking instead place, say, in the *real* Cartesian uv-plane via the position $x_n(t) \equiv u_n(t) + \mathbf{i}\, v_n(t)$ where u_n and v_n are *real* numbers and $\mathbf{i}$ is the *imaginary* unit, $\mathbf{i}^2 = -1$.

These N-body problems are called "solvable" if their initial-value problems— i.e., the evaluation of the N coordinates $x_n(t)$ at time t from given initial data $x_\ell(0)$ and $\dot{x}_\ell(0)$—can be performed by *algebraic* operations such as the identification of the N zeros of a time-dependent polynomial, see (2.3), the N *coefficients* $y_m(t)$ of which evolve in time in a manner that is either *explicitly* known or can itself be identified by *algebraic* operations or by *quadratures*. These dynamical systems are generally characterized by a time-evolution that does *not* feature any *chaotic* aspects and which in fact might be remarkably *simple*, for instance *multi-periodic* or even *isochronous* (i.e. such that *all solutions* are *completely periodic* with a *fixed* period—independent of the *initial* data of the problem—at least for an *open* set of these *initial* data).

We use the more specific additional term "integrable" for *solvable N*-body problems characterized by Newtonian equations of motion which can be shown to be *Hamiltonian*—and that, thanks to their *solvable* character, generally feature the existence of *N* independent constants of motion in involution, implying *Liouville integrability*.

While the formulas relevant for time-derivatives of higher order than 2 (see Subsection 2.2.2) are also relevant to some of the results reported below, the key formula to identify most of the newly identified *solvable/integrable* models reported as examples in the following sections of this Chapter 4 is (2.8), which is reprinted here for the convenience of the reader:

$$\ddot{x}_n = \sum_{\ell=1, \ell \neq n}^{N} \left(\frac{2 \dot{x}_n \dot{x}_\ell}{x_n - x_\ell} \right) - \left[\prod_{\ell=1, \ell \neq n}^{N} (x_n - x_\ell) \right]^{-1} \sum_{m=1}^{N} \left[\ddot{y}_m (x_n)^{N-m} \right]. \quad (4.79)$$

The strategy to identify new *solvable/integrable* models suggested by this formula is obvious: suppose that the quantities $y_m(t)$ evolve in time according to a set of *N* Newtonian equations of motion,

$$\ddot{y}_m = F_m\left(\vec{y}, \dot{\vec{y}}; t \right), \quad (4.80)$$

which is itself *solvable/integrable*. The *new* set of *N* Newtonian equations of motion which obtain by inserting these expressions of $\ddot{y}_m(t)$, (see (4.80)), in the right-hand side of (4.79), which read as follows,

$$\ddot{x}_n = \sum_{\ell=1, \ell \neq n}^{N} \left(\frac{2 \dot{x}_n \dot{x}_\ell}{x_n - x_\ell} \right)$$

$$- \left[\prod_{\ell=1, \ell \neq n}^{N} (x_n - x_\ell) \right]^{-1} \sum_{m=1}^{N} \left[F_m\left(\vec{y}, \dot{\vec{y}}; t \right) (x_n)^{N-m} \right], \quad (4.81)$$

are then as well *solvable/integrable*. Indeed their initial-value problem can be solved by first solving the corresponding initial-value problem for the *solvable/integrable* set of *N* ODEs (4.80), and by then identifying the *N* coordinates $x_n(t)$ as the *N zeros* of the monic polynomial with *coefficients* $y_m(t)$ (see (2.3)).

Let us emphasize that these equations (4.81) are a system of *N nonlinear second-order* evolution ODEs of Newtonian type for the *N particle coordinates* $x_n(t)$, because the coordinates $y_m(t)$ and their time-derivatives $\dot{y}_m(t)$ appearing in their right-hand sides are supposed to be replaced there by their *explicit* expressions in terms of the coordinates $x_n(t)$ and their time-derivatives $\dot{x}_n(t)$ via the formulas (2.4) and (2.5).

Remark 4.1. This system to generate *new solvable* systems of Newtonian equations of motion, see (4.81), from known systems of Newtonian equations of motion, see (4.80), can be iterated over and over again by using the models obtained at one level of the iteration in order to generate via this technique models at the next level of the iteration. In this manner *infinite hierarchies* of *solvable/integrable* Newtonian equations of motions are generated. We discuss in detail this possibility below, see Chapter 6. ∎

Remark 4.2. The N-body problem characterized by the system of Newtonian equations of motions (4.81) inherits many properties possessed by the, assumedly *solvable*, system of ODEs (4.80): for instance if the systems (4.80) are *isochronous* with periods $k_m T$ where the numbers k_m are *all positive nonvanishing integers*, implying that the corresponding polynomial (2.3) is itself periodic with period $K T$ where K is the Minimum Common Multiple of the N integers k_m, then the same property is as well featured by the system (4.81), albeit generally with a period that is itself a positive *integer multiple* ν of $K T$. (For a discussion of this important question—relevant to many of the N-body problems considered below—see Section 2.3). And another property that, if possessed by the N systems (4.80), is then automatically inherited by the system (4.81) is that of being *Hamiltonian*, since—in a *Hamiltonian* context—the transition from the *zeros* of a time-dependent polynomial to the *coefficients* of a time-dependent polynomial is a *point-transformation*, hence a *canonical transformation*. These various inheritance properties will be highlighted on a case-by-case basis in the following sections. ∎

In the following sections of this Chapter 4 we identify and investigate several such *solvable/integrable Newtonian*—and, in some cases, also *Hamiltonian*— N-body models. The presentation is drafted so as to facilitate the reader who is interested in the material reported in *only one* of the following sections or subsections of this Chapter 4.

4.1 The "Goldfish"

The simplest case is that in which the functions $F_m(\vec{y}, \dot{\vec{y}}; t)$, see (4.80) and (4.81), *vanish* identically; in this Section 4.1 we consider firstly the marginally more general case with $F_m(\vec{y}, \dot{\vec{y}}; t) = \mathbf{i}\, \omega\, \dot{y}_m$, so that (4.80) reads

$$\ddot{y}_m = \mathbf{i}\, \omega\, \dot{y}_m \,, \tag{4.82a}$$

implying the solution

$$y_m(t) = y_m(0) + \dot{y}_m(0) \left[\frac{\exp(\mathbf{i}\, \omega\, t) - 1}{\mathbf{i}\, \omega} \right]. \tag{4.82b}$$

Hereafter we assume the parameter ω to be a *real* number, so that each of the coefficients $y_m(t)$ is then periodic with period $T = 2\pi / |\omega|$ (provided $\omega \neq 0$).

It is then immediately seen—from (4.81) and (2.7)—that the corresponding system of evolution equations for the coordinates $x_n(t)$ reads as follows:

$$\ddot{x}_n = \mathbf{i}\,\omega\,\dot{x}_n + \sum_{\ell=1,\,\ell\neq n}^{N} \left(\frac{2\,\dot{x}_n\,\dot{x}_\ell}{x_n - x_\ell} \right). \qquad (4.83a)$$

Note the remarkable neatness of these equations of motion: they are *autonomous*, *invariant* under an *arbitrary common rescaling* and an *arbitrary common shift* of all the dependent variables x_n, and they only feature rather neat velocity-dependent two-body forces. Hence they imply that any particle initially at rest maintains this state of rest forever—since $\dot{x}_n = 0$ implies $\ddot{x}_n = 0$, see (4.83a)—and that its dynamics is then *completely decoupled* from that of the other particles.

It is also easily seen on the basis of the results described above that the solution of the initial-value problem for this N-body model is given by the following simple prescription: the values of the N coordinates $x_n(t)$ are given by the N roots of the following equation in the unknown z:

$$\sum_{n=1}^{N} \left(\frac{\dot{x}_n(0)}{z - x_n(0)} \right) = \frac{\mathbf{i}\,\omega}{\exp\,(\mathbf{i}\,\omega\,t) - 1} \qquad (4.83b)$$

(note that this is actually an algebraic equation of degree N in z, as it is immediately seen by multiplying it by the N-th degree polynomial $\prod_{n=1}^{N} [z - x_n(0)]$). This implies that, for $\omega \neq 0$, the model characterized by the equations of motion (4.83a) is *isochronous*, *all* its generic solutions being periodic with period $T = 2\pi / |\omega|$ or with an *integer* multiple ν of T due to the possibility that the N *zeros* of a time-dependent polynomial itself periodic in T be themselves periodic with a period νT *integer multiple* of T caused by an *exchange* of the positions of the *zeros* after a period (see Section 2.3).

The case with $\omega = 0$ features instead *generic* solutions with, typically, in the *remote past* $N-k$ particles *almost at rest* and k particles incoming from far away with *constant* velocity (of course $1 \leq k \leq N - 1$), and in the *remote future* an analogous configuration of $N-k$ particles *almost at rest* and k outgoing particles which inherited the speed of the incoming particles—with the possibility that particles exchange their positions and roles from the past to the future (so that this model, for $k = 1$, can be considered as mimicking a round of the parlor "game of musical chairs": see for instance Section 4.2.4 of [20]).

These findings are several decades old (see Section 4.N). They justify denoting as "models of goldfish type" *all* N-body problems characterized by Newtonian ("accelerations equal forces") equations of motion featuring on their ("forces") right-hand side the velocity-dependent two-body term $\sum_{\ell=1,\ \ell\neq n}^{N}\left[2\dot{x}_n\dot{x}_\ell/\left(x_n-x_\ell\right)\right]$.

Example 4.1.1. One such model which is a simple generalization of the standard goldfish model obtains starting from the following simple generalization of the equations of motion (4.82a):

$$\ddot{y}_m = \mathbf{i}\,\omega_m\,\dot{y}_m, \qquad (4.84a)$$

implying

$$y_m(t) = y_m(0) + \dot{y}_m(0)\left[\frac{\exp\left(\mathbf{i}\,\omega_m\,t\right)-1}{\mathbf{i}\,\omega_m}\right]. \qquad (4.84b)$$

Here we assume again the N parameters ω_m to be *all real*, so that each of the coefficients $y_m(t)$ is then periodic with period $T_m = 2\pi/|\omega_m|$ (provided $\omega_m \neq 0$); but we do *not* require any more that these parameters ω_m be *all equal*. The equations of motion for the N particle coordinates $x_n(t)$ then read as follows:

$$\ddot{x}_n = \sum_{\ell=1,\ \ell\neq n}^{N}\left(\frac{2\,\dot{x}_n\,\dot{x}_\ell}{x_n-x_\ell}\right) - \mathbf{i}\left[\prod_{\ell=1,\ \ell\neq n}^{N}\left(x_n-x_\ell\right)\right]^{-1}\sum_{m=1}^{N}\left[\omega_m\,\dot{y}_m\,(x_n)^{N-m}\right], \quad (4.84c)$$

with the N quantities $\dot{y}_m(t)$ expressed in terms of the N coordinates $x_n(t)$ and their time derivatives $\dot{x}_n(t)$ via the explicit formulas (2.5). The solutions of these equations of motion (4.84c) are provided by the N *zeros* $x_n(t)$ of the polynomial (2.3) where the N *coefficients* $y_m(t)$ are given by the formulas (4.84b) with $y_m(0)$ respectively $\dot{y}_m(0)$ given in terms of $x_n(0)$ and $\dot{x}_n(0)$ by the explicit formulas (2.4) respectively (2.5). Note that this implies that—if all the parameters ω_m are *real* and *nonvanishing*—then the *generic* motion in the *complex* x-plane of this N-body problem is *confined* and *multiply periodic*; it is moreover *isochronous* if the N numbers ω_m are all *nonvanishing* (*positive* or *negative*) *rational multiples* of the same (*nonvanishing*, *real*) number ω (i.e., $\omega_n = r_n\omega$, with the N parameters r_n being N arbitrary nonvanishing real *rational* numbers). ∎

Let us end this Section 4.1 with a mention of the *Hamiltonian* and *integrable* character of these equations of motion (4.84c). That this is the case is implied by the general argument provided above, see Remark 4.2, since clearly the model characterized by the equations of motion (4.84a) is itself *Hamiltonian* and *integrable*, these equations of motion being clearly yielded—as the diligent reader will easily verify—by the *Hamiltonian*

$$H\left(\vec{p};\,\vec{y}\right) = \sum_{m=1}^{N}\left[h_m\left(p_m;\,y_m\right)\right], \qquad (4.85a)$$

$$h_m\left(p;\,y\right) = -\mathbf{i}\left(\frac{\omega_m}{c_m}\right)y + \exp\left(c_m\,p\right)f_m\left(y\right), \qquad (4.85b)$$

where the parameters ω_m and c_m are *arbitrarily assigned* (of course *nonvanishing*), the functions $f_m(y)$ are also *arbitrarily assigned*, and the N variables $p_m(t)$ denote here the N *canonical momenta* conjugated to the N *canonical coordinates* $y_m(t)$. It is also obvious that this Hamiltonian is *integrable*, since the quantities $h_m(p_m; y_m)$ are N integrals of motion in involution.

4.2 Integrable Hamiltonian N-Body Problems of Goldfish Type Featuring N Arbitrary Functions

In Section 4.2 we treat a recently identified class of N-body problems of goldfish type, the remarkable character of which is advertised in the title of this Section 4.2; and we then discuss some specific examples in its subsections.

The starting point to arrive at this model is the set of N *decoupled* second-order ODEs of Newtonian type reading

$$\ddot{y}_m = f_m(y_m) \tag{4.86a}$$

which we assume to be satisfied by the coefficients $y_m(t)$ of a time-dependent polynomial, see (2.3). It is an elementary finding of classical mechanics that the initial-value problem of this system of *decoupled* ODEs can be solved by *quadratures*; and that this set of ODEs—featuring the N *a priori arbitrary* functions $f_m(y)$—is *Hamiltonian*, being equivalent to the set of Hamiltonian equations

$$\dot{y}_m = p_m, \quad \dot{p}_m = f_m(y_m), \tag{4.86b}$$

where the N coordinates $p_m \equiv p_m(t)$ are, in the Hamiltonian context, the *canonical momenta* conjugated to the *canonical coordinates* $y_m \equiv y_m(t)$.

The insertion of the set of ODEs (4.86a) in the basic identity (4.79) yields the following set of Newtonian equations of motions:

$$\ddot{x}_n = \sum_{\ell=1,\,\ell\neq n}^{N} \left(\frac{2\,\dot{x}_n\,\dot{x}_\ell}{x_n - x_\ell} \right) - \left[\prod_{\ell=1,\,\ell\neq n}^{N} (x_n - x_\ell) \right]^{-1} \sum_{m=1}^{N} \left[f_m(y_m)\ (x_n)^{N-m} \right],$$
$$\tag{4.87a}$$

with the quantities $y_m \equiv y_m(t)$ in the right-hand side replaced by their expressions (2.4) in terms of the coordinates $x_\ell \equiv x_\ell(t)$. These are then the N *Newtonian* equations of the N-body problem treated in Section 4.2, and it is clear from their derivation that this N-body problem features all the properties advertised in the title of this section, and that it is moreover *solvable*—up to *algebraic* operations—by *quadratures* (which in some cases can be *explicitly* performed in terms of *elementary* functions: see below).

For $N = 2$ this system, (4.87a), of 2 coupled nonlinear Newtonian equations of motion *of goldfish type* reads as follows:

$$\ddot{x}_n = (-1)^{n+1} (x_1 - x_2)^{-1} [2 \dot{x}_1 \dot{x}_2 - f_1 (-x_1 - x_2) x_n - f_2 (x_1 x_2)],$$
$$n = 1, 2. \tag{4.87b}$$

For $N = 3$ this system, (4.87a), of 3 coupled nonlinear Newtonian equations of motion *of goldfish type* reads as follows:

$$\ddot{x}_n = \sum_{\ell=1, \ \ell \neq n}^{N} \left(\frac{2 \dot{x}_n \dot{x}_\ell}{x_n - x_\ell} \right)$$

$$- \left[\prod_{\ell=1, \ \ell \neq n}^{3} (x_n - x_\ell) \right]^{-1} [f_1 (-x_1 - x_2 - x_3) (x_n)^2$$

$$+ f_2 (x_1 x_2 + x_2 x_3 + x_3 x_1) x_n + f_3 (-x_1 x_2 x_3)], \quad n = 1, 2, 3. \tag{4.87c}$$

Remark 4.2.1. If it is assumed that the N *arbitrary* functions $f_m(y)$ are *entire*, then clearly the only possible source of singularity in the evolution of the model of goldfish type (4.87a)—*additional* to those that might characterize the ODEs (4.86a)—are *particle collisions* i.e. the possibility that, at some time t_c during the time-evolution, two or more particle coordinates coincide, say $x_n (t_c) = x_\ell (t_c)$ with $n \neq \ell$, implying a blow-up of the right-hand side of (4.87a). This is *not* a generic event for motions of this N-body model of goldfish type taking place in the *complex* x-plane and starting from a generic set of *complex* initial data $x_n(0)$, $\dot{x}_n(0)$; while it is likely to happen for motions taking place on the *real* axis of the *complex* plane—as it might be the case when the functions $f_m(y)$ are *all real* functions of *real* arguments and the set of initial data $x_n(0)$ and $\dot{x}_n(0)$ include only *real* numbers (implying that the corresponding initial data $y_m(0)$ and $\dot{y}_m(0)$ are also *all real*, see (2.4) and (2.5)). Note that such collisions generally imply that the *identities* of the colliding particles after the collision are *undetermined*; and generally in the case of motions taking place, before the collision, on the *real* axis, the collision may cause the colliding particles to jump *off* the *real* axis, into the *complex* plane; in (clearly related) directions, since the coordinates of the two particles after the collision continue to be the two *zeros* of a *real* polynomial, hence two *complex conjugate* numbers. ∎

In some cases it is possible to get almost by inspection additional information on the behavior of the solutions of the N-body model (4.87a). For instance the following class of examples identify a subclass of N-body problems of goldfish type—see (4.87a)—featuring the property that *all* the corresponding motions starting from *arbitrary real* or *complex conjugate* initial data $x_n(0)$, $\dot{x}_n(0)$ are *confined* and *multiply periodic*. Let us indeed assume that the N functions $f_m(y)$ read as follows,

$$f_m(y) = (\bar{y}_m - y) \exp[g_m(y)], \tag{4.88}$$

where the N parameters $\bar{y}_m$ are N *arbitrary real* numbers and the N functions $g_m(y)$ are N *arbitrary real entire* functions of their argument y, and moreover such that $g_m(\pm\infty) \geq 0$. It is then evident—see (4.86a)—that the time evolution of the quantities $y_m(t)$ corresponding to *any* assignment of *real* initial data $y_m(0)$, $\dot{y}_m(0)$ takes place on the *real* axis of the *complex* y-plane and consists of periodic oscillations around the *equilibria* $y_m = \bar{y}_m$ with some periods T_m (generally depending on the initial data and on the functions (4.88)):

$$y_m(t + T_m) = y_m(t), \quad \dot{y}_m(t + T_m) = \dot{y}_m(t). \tag{4.89}$$

Hence—focussing now on the corresponding N-body model of goldfish type (4.87a) with (4.88)—for *any* assignment of initial data $x_n(0)$, $\dot{x}_n(0)$ of the N particle coordinates which only include *real* numbers or *complex conjugate* pairs so that the corresponding *initial data* $y_m(0)$ and $\dot{y}_m(0)$—see (2.4) and (2.5)—are *all real,* clearly the corresponding time evolutions of the coordinates $x_n(t)$ are *confined* and *multiply periodic*, because these coordinates are then the N *zeros* of a time-dependent polynomial, see (2.3), which is itself *multiply periodic*. These evolutions may take place on the *real* axis or in the *complex* plane (but then always featuring *pairs* of particles with *complex conjugate* coordinates), and may or may not feature *particle collisions*, see Remark 4.2.1. It is also evident that there are *open* sets of *real* initial data $x_n(0)$, $\dot{x}_n(0)$ yielding *nonsingular* motions with *no* particle collisions and the N particles remaining for *all* time confined to the *real* axis while the corresponding *real* coordinates $y_m(t)$ remain for *all* time each sufficiently close to their equilibrium values $\bar{y}_m$.

Let us complete this Section 4.2 with some examples, reported in the following subsections.

4.2.1 A Very Simple Isochronous Case

Consider the N-body problem characterized by the N ODEs (4.87a) with

$$f_m(y) = -(\omega_m)^2 \, y, \tag{4.90a}$$

hence reading

$$\ddot{x}_n = \sum_{\ell=1, \, \ell\neq n}^{N} \left(\frac{2\,\dot{x}_n\,\dot{x}_\ell}{x_n - x_\ell} \right) + \left[\prod_{\ell=1, \, \ell\neq n}^{N} (x_n - x_\ell) \right]^{-1} \sum_{m=1}^{N} \left[(\omega_m)^2 \, y_m \, (x_n)^{N-m} \right],$$
$$\tag{4.90b}$$

with the N quantities $y_m \equiv y_m(t)$ expressed in terms of the N coordinates $x_n \equiv x_n(t)$ via (2.4). The solution of the *initial-value* problem of this N-body model is then provided by the N zeros $x_n(t)$ of the following polynomial:

$$p_N \left(z; \, \vec{y}(t), \, \tilde{x}(t) \right) = z^N + \sum_{m=1}^{N} \left[y_m(t) \, z^{N-m} \right] = \prod_{n=1}^{N} \left[z - x_n(t) \right], \qquad (4.90c)$$

with

$$y_m(t) = y_m(0) \, \cos(\omega_m t) + \left[\frac{\dot{y}_m(0)}{\omega_m} \right] \sin(\omega_m t), \qquad (4.90d)$$

where the quantities $y_m(0)$ are expressed in terms of the initial values $x_n(0)$ by the standard formulas (2.4) and the quantities $\dot{y}_m(0)$ are expressed in terms of the initial values $x_n(0)$ and $\dot{x}_n(0)$ by the corresponding formulas (2.5). If all the parameters ω_m are *nonvanishing real* numbers this solution is confined to the *finite* part of the *complex x*-plane, being *multiply periodic* there, and the model is actually *isochronous* if the *nonvanishing real* numbers ω_m are *all* (*positive* or *negative*) *rational multiples* ρ_m of the same (*nonvanishing*) *real* parameter ω, $\omega_m = \rho_m \omega$.

> **Remark 4.2.1.1.** In the special case $\omega_m = \omega$, the system of N coupled nonlinear second-order ODEs (4.90b) becomes, via (2.6), the following neat system of N Newtonian equations of motion of goldfish type:
>
> $$\ddot{x}_n = \sum_{\ell=1, \, \ell \neq n}^{N} \left(\frac{2 \, \dot{x}_n \dot{x}_\ell}{x_n - x_\ell} \right) - \omega^2 \, (x_n)^N \left[\prod_{\ell=1, \, \ell \neq n}^{N} (x_n - x_\ell) \right]^{-1}, \qquad (4.90e)$$
>
> the solution of which is, of course, again provided by the same findings detailed above, now with $\omega_m = \omega$ (so that this model is obviously *isochronous*). ■

Let us complete this Subsection 4.2.1 by displaying the equations of motion (4.90b) in the simplest case with $N = 2$:

$$\ddot{x}_n = \frac{(-1)^{n+1} \left[2 \, \dot{x}_1 \dot{x}_2 - (\omega_1)^2 \, (x_1 + x_2) \, x_n + (\omega_2)^2 \, x_1 x_2 \right]}{x_1 - x_2}, \quad n = 1, 2 \, ; \quad (4.91)$$

and note that, via the transformation from the 2 *complex* coordinates x_n to the 2 *real* 2-vectors $\vec{r}_n \equiv (\operatorname{Re}[x_n], \, \operatorname{Im}[x_n])$ (see Appendix), they become—via the position $\omega_n = \rho_n \omega$—the equations (1.1) of Example 1.1 in Chapter 1.

4.2.2 A Case That Involves Elliptic Functions in the Solutions

Consider the N-body problem characterized by the N ODEs (4.87a) with

$$f_m(y) = -\delta(\mu_m) \, y_m \left[A_m - 2 \, C_m \, (y_m)^2 \right], \qquad (4.92a)$$

where $\delta(\mu)$ *vanishes* if μ vanishes and is *unity* otherwise ($\delta(\mu) = 0$ if $\mu = 0$, $\delta(\mu) = 1$ if $\mu \neq 0$), A_m and C_m are *a priori arbitrary* parameters, and the N parameters μ_m are arbitrarily assigned (if some of them are assigned a *vanishing* value, then the corresponding functions f_m should be identically set to *zero*; this being guaranteed by the factor $\delta(\mu_m)$, see (4.92a)). Then the solutions of the *initial-value* problem for the ODEs (4.86a) read

$$\ddot{y}_m = -\delta(\mu_m) \ y_m \left[A_m - 2 \ C_m \ (y_m)^2\right], \tag{4.92b}$$

and are simply $y_m(t) = y_m(0) + \dot{y}_m(0)t$ whenever the corresponding μ_m vanishes, $\mu_m = 0$; while when μ_m does *not* vanish, $\mu_m \neq 0$,

$$y_m(t) = a_m \ \text{sn} \left(b_m \ (t - \bar{t}_m) \, ; \ \kappa_m\right), \tag{4.93a}$$

$$(b_m)^2 = A_m \ \left(A_m - a_m^2 \ C_m\right), \tag{4.93b}$$

$$(\kappa_m)^2 = \left[1 - (a_m)^2 \ C_m\right]^{-1} - 1, \tag{4.93c}$$

where the function $\text{sn}(\tau; \kappa)$ is the standard Jacobian elliptic function (see for instance [72]) while the quantities a_m and $\bar{t}_m$ in (4.93a) can be evaluated in terms of the initial data $y_m(0)$ and $\dot{y}_m(0)$ and of the parameters A_m and C_m (see (4.93b) and (4.93c)).

With this assignment, (4.92a), of the functions $f_m(y)$, the equations of motion (4.87a) read

$$\ddot{x}_n = \sum_{\ell=1, \ \ell \neq n}^{N} \left(\frac{2 \ \dot{x}_n \ \dot{x}_\ell}{x_n - x_\ell}\right) + \left[\prod_{\ell=1, \ \ell \neq n}^{N} (x_n - x_\ell)\right]^{-1} \cdot$$
$$\cdot \sum_{m=1}^{N} \left\{\delta(\mu_m) \ y_m \left[A_m - 2 \ C_m \ (y_m)^2\right] \ (x_n)^{N-m}\right\}. \tag{4.93d}$$

These are the *Newtonian* equations of motion of an N-body model of goldfish type once the quantities y_m are expressed in terms of the particle coordinates x_n via the formulas (2.4). Then—if for all m in the interval from 1 to N—the parameters μ_m do not vanish, $\mu_m \neq 0$, clearly the *periodicity* of the Jacobian function $\text{sn}(\tau; \kappa)$ as function of τ (for $0 < \kappa^2 < 1$, see for instance [72]) guarantees that there is a set of initial data yielding *multiply periodic* solutions: clearly a sufficient condition is that the corresponding values of *all* the constants κ_m (see (4.93c)) be *real* and satisfy the inequalities $0 < (\kappa_m)^2 < 1$.

4.2.3 A Rather Simple Case and Its Isochronous Variant

Consider the N-body problem characterized by the N ODEs (4.87a) with

$$f_m(y) = 0 \quad \text{unless} \quad m = j, \ f_j(y) = c \ y^k, \tag{4.94a}$$

so that the equations of motion of the variables y_m read as follows (see (4.86a)):

$$\ddot{y}_m = \delta_{mj}\, c\, y^k. \tag{4.94b}$$

Here and throughout this Subsection 4.2.3 j is an *arbitrary integer* (in the range from 1 to N), k an *arbitrary integer* (*positive* or *negative*, but $k \neq 1$), and c is an *arbitrary* (possibly *complex*) parameter. Then the N ODEs (4.87a) read

$$\ddot{x}_n = \sum_{\ell=1,\,\ell\neq n}^{N}\left(\frac{2\,\dot{x}_n\,\dot{x}_\ell}{x_n - x_\ell}\right) - c\left[\prod_{\ell=1,\,\ell\neq n}^{N}(x_n - x_\ell)\right]^{-1}(y_j)^k\,(x_n)^{N-j}, \tag{4.94c}$$

with y_j expressed via the N coordinates x_n by (2.4).

The solution of this Newtonian N-body system of goldfish type is provided by the N zeros $x_n(t)$ of the polynomial (2.3) of degree N in z with, for $m \neq j$, $y_m(t) = y_m(0) + \dot{y}_m(0)t$, while $y_j(t)$ is the solution of the ODE $\ddot{y}_j = c(y_j)^k$, which is *solvable by quadratures*.

There are two different ways to "isochronize" this model, i.e. to manufacture new models featuring *multiperiodic* solutions, or possibly even *completely periodic* solutions emerging from an *open* set of initial data ("isochrony").

A first way is to apply directly to the system of nonlinear coupled equations (4.94c) the *trick* described above in detail in the context of a *single* nonlinear ODE: see the first part of Section 2.4. Let us indicate again here how this works.

The first step is to reformulate the equations of motion (4.94c) by first rewriting them via the purely notational change of (dependent and independent) variables $x_n(t) \Rightarrow \xi_n(\tau)$ (and correspondingly $y_m(t) \Rightarrow \gamma_m(\tau)$) so that (4.94c) read

$$\xi_n'' = \sum_{\ell=1,\,\ell\neq n}^{N}\left(\frac{2\,\xi_n'\,\xi_\ell'}{\xi_n - \xi_\ell}\right) - c\left[\prod_{\ell=1,\,\ell\neq n}^{N}(\xi_n - \xi_\ell)\right]^{-1}(\gamma_j)^k\,(\xi_n)^{N-j}, \tag{4.95}$$

where (above and hereafter) appended primes denote differentiations with respect to τ. Next, we introduce *new* x_n coordinates via the position

$$x_n(t) = \exp(\mathbf{i}\,r\,\omega\,t)\,\xi_n(\tau), \quad \tau \equiv \tau(t) = \frac{\exp(\mathbf{i}\,\omega\,t) - 1}{\mathbf{i}\,\omega}, \quad r = \frac{2}{(k-1)\,j} \tag{4.96a}$$

implying

$$\dot{x}_n(t) - \mathbf{i}\,r\,\omega\,x_n(t) = \exp[\mathbf{i}\,(r+1)\,\omega\,t]\,\xi_n'(\tau), \tag{4.96b}$$

$$\ddot{x}_n(t) - \mathbf{i}\,(2\,r+1)\,\omega\,\dot{x}_n(t) - r\,(r+1)\,\omega^2\,x_n(t) = \exp[\mathbf{i}\,(r+2)\,\omega\,t]\,\xi_n''(\tau), \tag{4.96c}$$

and the new quantity

$$y_j(t) = \exp\left(\mathbf{i}\,r\,j\,\omega\,t\right)\,\gamma_j(\tau), \tau \equiv \tau(t) = \frac{\exp\left(\mathbf{i}\,\omega\,t\right) - 1}{\mathbf{i}\,\omega}, \tag{4.96d}$$

that is related to the quantities x_n by the relation (2.4) with $m = j$.

It is then easily seen that, to the *autonomous* system of ODEs (4.95), there corresponds the following, also *autonomous*, Newtonian equations of motion describing a new *N*-body model:

$$\ddot{x}_n = \mathbf{i}\,(2\,r + 1)\,\omega\,\dot{x}_n + r\,(r+1)\,\omega^2\,x_n$$

$$+ \sum_{\ell=1,\,\ell\neq n}^{N} \left[\frac{2\,(\dot{x}_n - \mathbf{i}\,r\,\omega\,x_n)\,(\dot{x}_\ell - \mathbf{i}\,r\,\omega\,x_\ell)}{x_n - x_\ell} \right]$$

$$- c\left[\prod_{\ell=1,\,\ell\neq n}^{N} (x_n - x_\ell) \right]^1 \left(y_j\right)^k \left(x_n\right)^{N-j}, \tag{4.97a}$$

with

$$r = \frac{2}{(k-1)\,j}, \tag{4.97b}$$

and ω an arbitrary *real* parameter ($\omega \neq 0$; for $\omega = 0$ this generalized *isochronous* model (4.97) clearly reduces to the *nonisochronous* original model (4.94c)). Here the coefficient $y_j \equiv y_j(t)$ is expressed via the *N* coordinates $x_n \equiv x_n(t)$ by the standard formula (2.4). Note that the *autonomous* (i.e., time-independent) character of these equations of motion, (4.97a), is a consequence of the definition of r_j (see (4.97b) and (4.96d)). And the *isochronous* character of the model (4.97a) is implied by (4.96d), together with the *solvable* character of the model (4.94c) (see, if need be, Section 2.3).

An alternative, quite different, approach is to "isochronize" the ODEs (4.94b) satisfied by the variables y_m by replacing these ODEs as follows:

$$\ddot{y}_m = \mathbf{i}\,(2\,r_m + 1)\,\omega_m\,\dot{y}_m + r_m\,(r_m + 1)\,(\omega_m)^2\,y_m \quad \text{for} \quad m \neq j\,;$$

$$\ddot{y}_j = \mathbf{i}\,(2\,r_j + 1)\,\omega\,\dot{y}_j + r_j\,(r_j + 1)\,\omega^2\,y_j + c\,\left(y_j\right)^k. \tag{4.98a}$$

Here and below, the $N - 1$ parameters r_m with $m \neq j$ are assumed to be *arbitrary real rational* numbers,

$$r_m = \frac{q_m}{k_m}, \quad m \neq j, \tag{4.98b}$$

with q_m and k_m arbitrary *coprime integers* ($k_m \geq 1$, $q_m \neq 0$), while

$$r_j = \frac{2}{k-1}\,; \tag{4.98c}$$

and the $N - 1$ parameters ω_m might be *arbitrary real* numbers, but we prefer to assume that they are *real rational* multiples of the parameter ω which is itself an *arbitrary real nonvanishing* number:

$$\omega_m = \rho_m \, \omega, \quad \rho_m = \frac{\lambda_m}{\mu_m}, \quad m \neq j, \tag{4.98d}$$

with λ_m and μ_m *arbitrary coprime integers* ($\mu_m \geq 1$, $\lambda_m \neq 0$).

The solution of these ODEs reads as follows: for $m \neq j$,

$$y_m(t) = y_m(0) + \left[\dot{y}_m(0) - \mathbf{i}\, r_m \, \omega_m \, y_m(0)\right] \tau_m(t), \quad \tau_m(t) = \frac{\exp\left(\mathbf{i}\, \omega_m \, t\right) - 1}{\mathbf{i}\, \omega_m}, \tag{4.99a}$$

while, for $m = j$,

$$y_j(t) = \exp\left(\mathbf{i}\, r_j \, \omega \, t\right) \gamma\left(\tau\right), \quad \tau = \frac{\exp\left(\mathbf{i}\, \omega \, t\right) - 1}{\mathbf{i}\, \omega}, \tag{4.99b}$$

with the function $\gamma\left(\tau\right)$ satisfying (see (4.94b)) the ODE

$$\gamma'' = c\, \gamma^k. \tag{4.99c}$$

This last ODE can be solved by *quadratures* for any value of the parameter k, but the corresponding solution is generally *not* an elementary function. Here we restrict attention to the case with $k = -3$, when $r_j = -1/2$ (see 4.98c)) and the solution reads simply as follows:

$$\gamma\left(\tau\right) = \gamma\left(0\right) \left[1 + \frac{2\, \gamma'(0)\, \tau}{\gamma\left(0\right)} + a\, \tau^2\right]^{1/2},$$

$$a = c\, \left[\gamma\left(0\right)\right]^{-2} \left\{1 + c^{-1}\, \left[\gamma\left(0\right) \gamma'(0)\right]^2\right\}. \tag{4.99d}$$

Hence in this case the Newtonian equations of motion of the N-body problem satisfied by the coordinates $x_n(t)$ reads as follows:

$$\ddot{x}_n = \sum_{\ell=1,\,\ell\neq n}^{N} \left(\frac{2\, \dot{x}_n \, \dot{x}_\ell}{x_n - x_\ell}\right) - \left[\prod_{\ell=1,\,\ell\neq n}^{N} (x_n - x_\ell)\right]^{-1} \cdot$$

$$\cdot \left\{ \sum_{m=1,\,m\neq j}^{N} \left\{\left[\mathbf{i}\, (2\, r_m + 1)\, \rho_m \, \omega \, \dot{y}_m + r_m\, (r_m + 1)\, (\rho_m \, \omega)^2\, y_m\right] (x_n)^{N-m}\right\} \right.$$

$$\left. + c\, \left(y_j\right)^{-3}\, (x_n)^{N-j} \right\}. \tag{4.100}$$

Here the parameter c is an *arbitrary complex* number, the parameter ω is an *arbitrary nonvanishing real* number, and the $2\,(N - 1)$ parameters r_m and ρ_m with

$m \neq j$ are *arbitrary nonvanishing rational numbers*, while $r_j = -1/2$ and $\rho_j = 1$, and the $2N$ quantities $y_m \equiv y_m(t)$ respectively $\dot{y}_m \equiv \dot{y}_m(t)$ are all expressed via the N coordinates $x_n \equiv x_n(t)$ and $\dot{x}_n \equiv \dot{x}_n(t)$ by the standard formulas (2.4) respectively (2.5).

The solution of these equations of motion (4.100) are then provided by the N *zeros* $x_n(t)$ of the polynomial (2.3) the coefficients $y_m(t)$ of which evolve as detailed by the formulas (4.99) with, in (4.99d), $\gamma(0) = y_j(0)$ and $\gamma'(0) = \dot{y}_j(0) + \mathbf{i}(\omega/2)\, y_j(0)$. In all these formulas *all* the *quantities* $y_m(0)$ and $\dot{y}_m(0)$ must be replaced by their expressions (2.4) and (2.5) in terms of the *initial* data $x_n(0)$ and $\dot{x}_n(0)$.

The *isochronous* character of this N-body problem is a consequence of the periodicity of the polynomial (2.3) implied by these findings, see (4.99) with (4.99d) (and, if need be, Section 2.3).

Let us end this Section 4.2.3 by displaying the equations of motion (4.100) in the simplest $N = j = 2$ case (implying $r_2 = -1/2$, $r_2 = 1$):

$$\ddot{x}_n = \frac{(-1)^{n+1}}{x_1 - x_2} \left\{ 2\, \dot{x}_1\, \dot{x}_2 - c\, (x_1\, x_2)^{-3} \right.$$
$$\left. + \left[\mathbf{i}\,(2\,r_1 + 1)\,\rho_1\,\omega\,(\dot{x}_1 + \dot{x}_2) + r_1\,(r_1 + 1)\,(\rho_1\,\omega)^2\,(x_1 + x_2)\right] x_n \right\}.$$
$$(4.101)$$

Note that this equation becomes real if also $r_1 = -1/2$.

4.2.4 Another Rather Simple Case, Its Isochronous Variant, and Its Treatment in the Real Context

The fourth N-body problem of goldfish type we consider in Section 4.2 is the *integrable Hamiltonian* model characterized by the system of nonlinear ODEs

$$\ddot{x}_n = \sum_{\ell=1,\,\ell\neq n}^{N} \left(\frac{2\,\dot{x}_n\,\dot{x}_\ell}{x_n - x_\ell}\right) + \left[\prod_{\ell=1,\,\ell\neq n}^{N} (x_n - x_\ell)\right]^{-1} \cdot$$
$$\cdot \sum_{m=1}^{N} \left\{ (\omega_m)^2 \left[y_m - (\lambda_m)^2\,(y_m)^{-3} \right] (x_n)^{N-m} \right\}, \qquad (4.102a)$$

corresponding to (4.87a) with

$$f_m(y) = (\omega_m)^2 \left[-y + (\lambda_m)^2\,y^{-3} \right]. \qquad (4.102b)$$

The ODEs (4.102a) are a *nonlinear coupled* system of N Newtonian equations of motion of goldfish type for the N particle coordinates $x_n(t)$ inasmuch as the quantities $y_m(t)$ appearing in their right-hand sides are supposed to be replaced by their *explicit* expressions (2.4) in terms of these particle coordinates.

Here the parameters ω_m and λ_m are *2N arbitrary real* numbers (note that this model is a generalization of that treated above, see Subsection 4.2.1, to which it reduces for $\lambda_m = 0$).

In this book we generally consider the motion of *N*-body problems in the *complex* plane, and the treatment as given below is valid in that context (up to obvious minor changes); but in Subsection 4.2.4 we entertain the interested reader in a detailed discussion focused on motions taking instead place on the *real* line, to provide some feeling for the additional complications entailed by such a restriction.

The solution of this *N*-body problem (4.102a) is provided by the *N zeros* $x_n(t)$ of the polynomial (2.3) the *coefficients* $y_m(t)$ of which evolve according to the system of *nonlinear decoupled* ODEs

$$\ddot{y}_m = (\omega_m)^2 \left[-y_m + (\lambda_m)^2 \ (y_m)^{-3} \right]. \tag{4.102c}$$

And it is easily seen that the explicit solution of the initial-value problem of these ODEs reads as follows:

$$y_m(t) = a_m \left\{ \sin^2 \left[\omega_m \ (t - t_m) \right] + c_m \right\}^{1/2}, \tag{4.103a}$$

$$a_m = \left(h_m^2 - 4 \lambda_m^2 \right)^{1/4}, \quad c_m = \frac{1}{2} \left(\frac{h_m}{a_m^2} - 1 \right), \tag{4.103b}$$

$$h_m = \left[\frac{\dot{y}_m(0)}{\omega_m} \right]^2 + \left[y_m(0) \right]^2 + (\lambda_m)^2 \ \left[y_m(0) \right]^{-2}, \tag{4.103c}$$

$$t_m = \omega_m^{-1} \ \arcsin \left(\left\{ \left[\frac{y_m(0)}{a_m} \right]^2 - c_m \right\}^{1/2} \right), \tag{4.103d}$$

where (see (2.4) and (2.5))

$$y_m(0) = (-1)^m \sum_{1 \leq n_1 < n_2 < \cdots < n_m \leq N} \left[x_{n_1}(0) \cdot x_{n_2}(0) \cdots x_{n_m}(0) \right], \tag{4.103e}$$

$$\dot{y}_1(0) = - \sum_{n=1}^{N} \left[\dot{x}_n(0) \right], \tag{4.103f}$$

$$\dot{y}_m(0) = (-1)^m \sum_{n=1}^{N} \left\{ \dot{x}_n(0) \sum_{\substack{1 \leq n_1 < n_2 < \cdots < n_m \leq N, \\ n_j \neq n, \ j=1,2,\ldots,m}} \left[x_{n_1}(0) \cdot x_{n_2}(0) \cdots x_{n_{m-1}}(0) \right] \right\},$$

$$m = 2, 3, \ldots, N. \tag{4.103g}$$

The coefficients $y_m(t)$ are defined by *continuity* in t from their initial values $y_m(0)$ (i.e., this prescription identifies the determination of the roots appearing in the above formulas, see (4.103)); and note that if $\omega_m = 0$ and $\omega_m \lambda_m \equiv \mu_m$ also vanishes, $\mu_m = 0$, the formula (4.103a) is replaced by

$$y_m(t) = y_m(0) + \dot{y}_m(0)\, t\,; \qquad (4.103\text{h})$$

while if $\omega_m > 0$ and $\lambda_m = \mu_m = 0$ (implying $c_m = 0$, see (4.103b)), then

$$y_m(t) = y_m(0)\, \cos(\omega_m t) + \frac{\dot{y}_m(0)}{\omega_m}\, \sin(\omega_m t) = A_m \cos(\omega_m t + \theta_m)\,,$$

$$A_m = \left\{ \left[y_m(0) \right]^2 + \left[\frac{\dot{y}_m(0)}{\omega_m} \right]^2 \right\}^{1/2}\,, \qquad \tan(\theta_m) = -\frac{\dot{y}_m(0)}{\omega_m\, y_m(0)}\,; \qquad (4.103\text{i})$$

and if $\omega_m = 0$ but $\omega_m \lambda_m \equiv \mu_m$ is a *nonvanishing real* number (for definiteness, *positive*: $\mu_m > 0$), then

$$y_m(t) = y_m(0)\, \left\{ \rho_m^2 + \left[\left(1 - \rho_m^2 \right)^{1/2} + \frac{\mu_m t}{\rho_m \left[y_m(0) \right]^2} \right]^2 \right\}^{1/2}\,,$$

$$\rho_m = \mu_m \left| \left\{ \mu_m^2 + \left[y_m(0)\, \dot{y}_m(0) \right]^2 \right\}^{-1/2} \right|\,, \qquad 0 < \rho_m \leq 1, \qquad (4.103\text{j})$$

implying, as $t \to \pm\infty$,

$$y_m(t) = \left| \frac{\mu_m t}{\rho_m\, y_m(0)} \right| + y_m(0) \left(1 - \rho_m^2 \right)^{1/2} + O\left(\left| \frac{1}{t} \right| \right)\,. \qquad (4.103\text{k})$$

Let us now discuss the properties of these solutions, (4.103), when they evolve from *real* initial data $x_n(0)$, $\dot{x}_n(0)$.

Remark 4.2.4.1. The very structure of the equations of motion (4.102a) requires that the initial data $x_n(0)$ be *all different among themselves*, i.e. $x_n(0) \neq x_\ell(0)$ if $n \neq \ell$, and moreover, if $\omega_m \lambda_m \equiv \mu_m \neq 0$, then the initial data must imply $y_m(0) \neq 0$ (see (4.102c)); hence, in particular, the condition $y_N(0) = (-1)^N x_1(0) x_2(0) \cdots x_N(0) \neq 0$, see (4.103e), must hold, implying $x_n(0) \neq 0$ for *all* values of the index n in its range from 1 to N. ∎

Let us now, first of all, restrict consideration to initial data for which the system of evolution equations (4.102a) *does not* run into a *singularity* at a *finite* time (as discussed below); as implied by the following discussion, these initial data are *generic*, namely they belong to an *open* set of initial data sharing the property to yield *nonsingular* solutions. Let us immediately note that—in the *nonsingular* case to which we restrict for the moment attention, with *all* the parameters ω_m and $\omega_m \lambda_m \equiv \mu_m$ *real* and (for definiteness) *nonnegative*, and the N coordinates $x_n(t)$, as well as the N quantities $y_m(t)$ are also *real*—each function $y_m(t)$ is clearly in all cases *nonsingular* (if $\omega_m > 0$ and $\omega_m \lambda_m \equiv \mu_m > 0$, see (4.103a) and note that (4.103b) with (4.103c) imply $c_m > -1$; and quite evidently in the other cases, see (4.103h), (4.103i) and (4.103j)). Moreover, each function $y_m(t)$ is clearly periodic if $\omega_m > 0$ and $\omega_m \lambda_m \equiv \mu_m > 0$ (see (4.103a)),

$$y_m(t + T_m) = y_m(t), \quad T_m = \pi/\omega_m \ \text{ if } \ \omega_m > 0, \quad \omega_m \lambda_m \equiv \mu_m > 0 , \quad (4.104a)$$

and it is also periodic if $\omega_m > 0$ and $\omega_m \lambda_m \equiv \mu_m = 0$, but then with period $2T_m$ (see (4.103i)),

$$y_m(t + 2\,T_m) = y_m(t), \quad T_m = \pi/\omega_m \ \text{ if } \ \omega_m > 0, \quad \omega_m \lambda_m \equiv \mu_m = 0 ; \quad (4.104b)$$

it is instead *not* periodic if $\omega_m = 0$ and λ_m is *finite* so that $\omega_m \lambda_m \equiv \mu_m = 0$ (see (4.103h)), and also if $\omega_m = 0$ but $\omega_m \lambda_m \equiv \mu_m > 0$, see the last three formulas of the set (4.103). It is also clear (see (4.103c), (4.103b) and (4.103a)) that if $\omega_m > 0$ and $\omega_m \lambda_m \equiv \mu_m > 0$ there are 2 *real equilibrium* (i.e., time-independent) values of $y_m(t)$,

$$y_m(t) = y_m(0) = \pm (\lambda_m)^{1/2} , \quad \dot{y}_m(t) = 0 , \qquad (4.105)$$

and if $\omega_m \geq 0$ and $\omega_m \lambda_m \equiv \mu_m = 0$ the *only* equilibrium value is $y_m(t) = y_m(0) = 0$, $\dot{y}_m(t) = 0$. While in the other case—if ω_m vanishes, but $\omega_m \lambda_m \equiv \mu_m > 0$—then the quantities $y_m(t)$ always *do* evolve in t (no equilibria), i.e., $|y_m(t)| + |\dot{y}_m(t)| > 0$ for all time (see (4.103j)), with their asymptotic behaviors in the remote past and future detailed by (4.103k).

Next, let us discuss how the solutions $x_n(t)$ of the Newtonian equations of motion (4.102a) actually behave. Let us firstly consider the *more generic* case with both $\omega_m > 0$ and $\omega_m \lambda_m \equiv \mu_m > 0$ for all values of the index m in its range from 1 to N, always focusing to begin with on the *nonsingular* case.

> **Remark 4.2.4.2.** The structure of the equations of motion (4.102a) implies that these *nonsingular* solutions are *all real*—since we do assume to start from *real* initial data $x_n(0)$, $\dot{x}_n(0)$—hence that the order of the N points $x_n(t)$ on the *real* line cannot change during the time evolution, since such a change would require that at some time one of the moving point overtakes another one of the moving points—i.e., that $x_n(t_c) = x_\ell(t_c)$ with $n \neq \ell$ at some time t_c—and clearly this would imply a blow-up of the right-hand side of the equations of motion (4.102a). For an analogous reason throughout the time evolution the N quantities $y_m(t)$ never vanish, $y_m(t) \neq 0$, if $\omega_m \lambda_m \equiv \mu_m > 0$.
>
> We show below that solutions having these properties *do exist*—at least provided *none* of the parameters ω_m and $\omega_m \lambda_m \equiv \mu_m$ vanishes ($\omega \neq 0$, $\mu_m \neq 0$; with the parameters λ_m possibly required to satisfy some *inequalities*, see below)—and indeed emerge from an *open* set of initial data $x_n(0)$, $\dot{x}_n(0)$. ∎

It is then clearly implied that for *any* set of initial data $x_n(0)$, $\dot{x}_n(0)$, the N coordinates $x_n(t)$ all remain in a *finite* region of the *real* axis—here the condition $\omega_m \neq 0$ is essential—and their time evolution is *multiply periodic*, since they are the *zeros* of a polynomial the time-dependence of which is itself *multiply periodic* (indeed, each of its coefficients $y_m(t)$ is *periodic* with its period T_m; see (4.104a)). It is moreover evident that, if the N parameters ω_m are all *rational multiples* of the same (*real, nonvanishing*) number ω, i.e.,

$$\omega_m = r_m\,\omega, \quad r_m = \frac{q_m}{k_m}, \quad m = 1, 2, \ldots, N\,, \tag{4.106a}$$

with q_m and k_m *coprime integers* ($k_m \geq 1$, $q_m \neq 0$), then the relevant polynomial is itself *completely periodic* with at most (see Remark 2.4.1.2) period

$$T = \frac{k\,\pi}{\omega}, \quad k = \underset{m=1,2,\ldots,N}{\text{MinimumCommonMultiple}}\,[k_m]\,; \tag{4.106b}$$

then *all nonsingular* solutions $x_n(t)$ are periodic as well, with the same period (see the following Remark 4.2.4.3),

$$x_n\,(t + T) = x_n(t)\,, \quad n = 1, 2, \ldots, N\,, \tag{4.106c}$$

namely in this case the *Newtonian N*-body problem characterized by the equations of motion (4.102a) is *isochronous* (i.e., it features an *open* set of initial data yielding solutions periodic with a *fixed* period T independent of the initial data).

> **Remark 4.2.4.3.** If a time-dependent polynomial $P_N(z; t)$ of degree N in its argument z is periodic in the time t—$P_N(z; t + T) = P_N(z; t)$—then obviously the *unordered* set of its N zeros $x_n(t)$ is as well periodic with the same period; while the *ordered* set of its N zeros $x_n(t)$ is also periodic, but possibly with a *larger* period (a *positive integer multiple* ν of T) because the zeros may exchange their roles over the time evolution (see Section 2.3). However, in the present case, *all* the N zeros $x_n(t)$ must maintain their ordering on the *real* line throughout their time evolution— see Remark 4.2.4.2—hence they are *de facto* an *ordered* set throughout their time evolution. This validates the above statement, see (4.106c). ∎

The doubt might still linger in the mind of the reader about the *existence* of any such *nonsingular* solution of the system of Newtonian equations of motion (4.102a). To dispel it let us consider the *equilibrium* configurations—$x_n(t) = \bar{x}_n$, $\dot{x}_n(t) = 0$—of this N-body model. They correspond to the *equilibrium* values of the coefficients y_m, see (4.105), i.e., the N coordinates $\bar{x}_n$ are the N zeros $\bar{x}_n \equiv \bar{x}_n\,(\vec{s})$ of one of the following 2^N polynomials:

$$\bar{P}_N\,(z; \vec{s}) = \prod_{n=1}^{N}\,[z - \bar{x}\,(\vec{s})] = z^N + \sum_{m=1}^{N}\,\big[s_m\,\big|(\lambda_m)^{1/2}\big|\,z^{N-m}\big], \quad s_m = \pm 1\,, \tag{4.107}$$

where the N-vectors $\vec{s}$ feature as their components s_m the values ± 1: so there are 2^N *different* assignments of these N-vectors $\vec{s}$, corresponding—for every assignment of the N parameters λ_m—to as many *different* polynomials $\bar{P}_N(z; \vec{s})$ (one-half of which are, however, identical to the other half up to an overall sign, and feature therefore the *same* set of zeros). So there are—for every assignment of the N parameters λ_m—at most 2^{N-1} *different* equilibrium configurations of the system of N Newtonian equations of motion (4.102c). But many of these configurations are not acceptable if we limit our attention—as we do in this discussion—to *real* values of the coordinates $x_n \equiv x_n(t)$: indeed a polynomial with *real* coefficients need not have *zeros* which are *all real* (it might also feature *pairs* of *complex conjugate zeros*). Moreover, an additional condition that we must require on the set of N zeros $\bar{x}_n$ in order that they qualify as an equilibrium configuration of the system of Newtonian equations (4.102a) is—see above—that they be *all different among themselves* and that *none* of their N symmetrical sums vanish. On the other hand there obviously are assignments of the N parameters λ_m and s_m that do imply that the N zeros $\bar{x}_n$ of the polynomial (4.107) feature *all* these properties: indeed one can clearly assign *any* set of N zeros $\bar{x}_n$ featuring these properties and then manufacture a polynomial $\bar{P}_N(z; \vec{s})$, see (4.107), featuring these numbers as its zeros by assigning the parameters λ_m and s_m as follows:

$$s_m \left| (\lambda_m)^{1/2} \right| = (-1)^m \sum_{1 \le n_1 < n_2 < \cdots < n_m \le N} \left(\bar{x}_{n_1} \cdot \bar{x}_{n_2} \cdots \bar{x}_{n_m} \right). \qquad (4.108a)$$

Note that the conditions mentioned above imply that the right-hand sides of these N equations *not vanish*, so that $\lambda_m > 0$ for all values of m in its range from 1 to N. Hence the system of Newtonian equations of motion (4.102a) with this assignment of the N parameters λ_m features a *nonsingular* time evolution for the following *open* set of initial data:

$$x_n(0) = \bar{x}_n + \varepsilon \, \alpha_n, \quad \dot{x}_n(0) = \varepsilon \, \beta_n, \qquad (4.108b)$$

for *any arbitrary* assignment of the $2N$ *real* numbers α_n and β_n, provided the parameter ε is *sufficiently small*; the corresponding solutions feature N coordinates $x_n(t)$ each of which oscillates according to a *multiply periodic* pattern around its equilibrium value $\bar{x}_n$—remaining sufficiently close to it not to give rise to any particle collision. These motions would be *completely periodic* in the *isochronous* case, see (4.106).

Qualitatively analogous results also hold if the restriction on all the parameters $\omega_m \lambda_m \equiv \mu_m$ to be *positive* is abandoned, allowing some or all of them to *vanish*; a detailed analysis of this more general case is left to the interested reader.

The case in which some of the parameters ω_m vanish—with $\omega_m \lambda_m$ also vanishing or having a *finite real* value μ_n—is instead *qualitatively* different, since then the corresponding coefficients $y_m(t)$ of the polynomial (2.3) do not remain all bounded for all time: indeed those corresponding to vanishing values of ω_m generally *diverge* in the remote past and future, see (4.103h) or the last 3 formulas in the set (4.103).

Let us discuss these two cases separately. So, let us firstly assume that for *some* (but *not all*) values of the index m in its range from 1 to N which we denote $\check{m}$ the corresponding value of ω_m vanish, $\omega_{\check{m}} = 0$, and the corresponding value of $\omega_{\check{m}}\lambda_{\check{m}} \equiv \mu_{\check{m}}$ also vanish, $\omega_{\check{m}}\lambda_{\check{m}} \equiv \mu_{\check{m}} = 0$, so that the time-evolution of $y_{\check{m}}(t)$ is given by the simple (linear) formula (4.103h). Hence the *generic nonsingular* solutions of the N-body problem characterized by the N Newtonian equations of motion (4.102a) feature at least one particle incoming in the remote past from far away—from the right or the left, as the case may be—and at least one particle escaping far away in the remote future—as implied by the requirement that, for the *nonsingular* solutions, the ordering of the particles on the real line cannot change throughout the time evolution (see Remark 4.2.4.2) and by the obvious fact that if one or more coefficients of a time-dependent polynomial *diverge* asymptotically as $t \to \pm\infty$, at least one *zero* of that polynomial must also *diverge* in that limit (for a more detailed treatment of this phenomenology see for instance Appendix G, entitled "Asymptotic behavior of the zeros of a polynomial whose coefficients diverge exponentially", of the book [20]). But there clearly also are *nongeneric* solutions—characterized by initial data such that for all values of $\check{m}$ the quantities $\dot{y}_{\check{m}}(0)$ vanish ($\dot{y}_{\check{m}}(0) = 0$, see (4.103f), (4.103g) and (4.103g))—that are *multiply periodic* (or possibly *completely periodic*, depending on the values of the other parameters ω_m, as discussed above, see (4.106)).

Finally, let us consider the case in which for some (but not all) values of the index m in its range from 1 to N which we denote $\hat{m}$ the corresponding values of ω_m vanish, $\omega_{\hat{m}} = 0$, but the corresponding values of $\omega_m\lambda_m \equiv \mu_m$ take the *finite* (*nonvanishing*) value $\mu_{\hat{m}}$, $\omega_{\hat{m}}\lambda_{\hat{m}} \equiv \mu_{\hat{m}} > 0$, so that the time evolution of $y_{\hat{m}}(t)$ is given by the last 3 formulas in the set (4.103). In this case clearly the same asymptotic behavior of the system as described just above will *always* prevail, namely the N-body problem does *not* feature any multiply periodic or periodic solution in which *all* the particles remain for *all* time in a *finite* region of the *real* axis: in the remote past and future at least one particle will come in from infinity and another one will escape to infinity.

More complete analyses of this phenomenology—including models featuring vanishing values of some parameters ω_m with associated values of the parameters $\omega_m\lambda_m \equiv \mu_m$, some of which also *vanish* while some others have a finite (*nonvanishing*) value—require a more detailed study, in particular when several particles come in from far away in the remote past, possibly some from the left and others from the right, and several escape far away in the remote future.

We end at this point our terse description of the phenomenological behavior of the *nonsingular* solutions of the N-body models characterized by the Newtonian equations of motion (4.102a), in the context of *real-valued* variables and parameters, and turn to an, also terse, description of the solutions of these N-body models that instead do run into a *singularity* at some finite time t_S. There clearly are two sources of singularity which are *a priori* evident from the very structure of the system of N coupled ODEs (4.102a), causing their right-hand side to blow up: *(i)* the eventual *vanishing* at some finite time $\hat{t}_m$ of one of the quantities $y_m(t), y_m(\hat{t}_m) = 0$, which is only relevant provided the corresponding parameter $\omega_m \lambda_m \equiv \mu_m$ does *not* vanish; *(ii)* the eventual "collision" of two (or possibly more) particles, namely the fact that at some time t_c two different particles meet, $x_n(t_c) = x_\ell(t_c)$ with $n \neq \ell$. And the solutions of the *nonlinear* system of ODEs (4.102a) might run into a singularity just because of its *nonlinear* character. However, both this last possibility and the first one mentioned just above are clearly *excluded*—in the *real-valued* context of all quantities to which we are restricting attention in Subsection 4.2.4—by the explicit solution of these models provided above by the N zeros of the polynomial (2.3) the *coefficients* of which are provided by the formulas (4.103). The second possibility cannot on the other hand be excluded, corresponding to the circumstance that, at the time t_c, the polynomial (2.3) with (4.103) feature a *double* (or perhaps *multiple*) *zero*. If we were working in a *complex* rather than *real* context the (*complex*) initial data yielding such an outcome would certainly be *nongeneric*, namely the *singular* outcome would disappear for any arbitrarily small but *generic* change of the initial data. But clearly this is *not* the case in our *real* context, when all the point-particles with *real* coordinates $x_n(t)$ are moving on the *one-dimensional real* axis of the *complex* x-plane, so that collisions are *a priori* likely to occur.

The typical singularity associated with a particle collision causes the coordinates of all particles to develop an *imaginary* part. If the N-body problem were describing the motion of particles in the *complex* x-plane, this phenomenon would be interpreted as a jumping off of the particles from the *real* axis into the *complex* plane; but note that in the process the colliding particles lose their identity. In our *real* context the occurrence of a *singularity* simply means a *breakdown* of the validity of the "physical" model, with no natural interpretation of what happens next. But let us reemphasize that, at least in the case with $(\omega_m)^2 > 0$ for all vales of m characterized by *confined* motions and the presence of *equilibria* there always is an *open* set of initial data yielding *nonsingular* solutions, as shown above. And it stands to reason that *open* sets of initial data yielding *nonsingular* solutions also occur when some of the parameters $(\omega_m)^2$ vanish implying that the generic ("scattering") solutions of the corresponding N-body problem characterized by the Newtonian equations of motion (4.102a) feature at least one particle moving, in the remote past and future, away from the *finite* part of the *real* line.

Let us also mention that a rather trivial extension of the model treated in this subsection, that allows the introduction of *N* *additional arbitrary* parameters which then enter in the equations of motion in a not quite trivial manner, is obtained by replacing in the Newtonian equations of motion (4.102a)—but not in the relations (4.103)—the quantities $y_m(t)$ with $y_m(t) - \eta_m$, where the *real* numbers η_m are *N* *arbitrary* parameters. The more general *N*-body model obtained in this manner has analogous properties to those reported above, which the interested reader will have no difficulty to uncover.

For completeness let us repeat that the set of ODEs (4.102c) is *Hamiltonian*, being yielded—as the reader will easily verify—by the Hamiltonian

$$H\,(\vec{p};\,\vec{y}) = \sum_{m=1}^{N} \left[H_m\,(p_m;\,y_m)\right], \quad H_m\,(p;\,y) = \frac{1}{2}\,\left\{p^2 + \omega_m^2\,\left[y^2 + (\lambda_m)^2\,y^{-2}\right]\right\},$$

(4.109)

where the *N* variables p_m are the *canonical momenta* conjugated to the *canonical coordinates* y_m. This Hamiltonian is obviously *integrable*, indeed it features the *N* constants of motion $H_m\,(p_m;\,y_m)$, which are clearly in involution. Moreover, the *N* canonical coordinates y_m are related to the *N* particle coordinates x_n by the formulas (2.4), which neither involve the canonical momenta p_m nor the time *t*: so, in Hamiltonian parlance, the change of variables from the *N* coordinates y_m to the *N* coordinates x_n—corresponding to the transition from the system of ODEs (4.102c) to the Newtonian equations of motion (4.102a)—is a *point transformation*, which is a special class of *canonical transformations*. Hence the time evolution (4.102a) of the *N* particle coordinates x_n is as well an *integrable Hamiltonian* system (note that we just repeated here the *inheritance* argument of Remark 4.2).

Let us end this Subsection 4.2.4 by reporting the equations of motion of this model, and by outlining the phenomenological aspects of their solutions, in the simplest case, that with $N = 2$. The quantities used below are always defined as above (but for the convenience of the reader we occasionally recall below their definitions).

The equations of motion (4.102a) with $N = 2$ read as follows:

$$\ddot{x}_n = (-1)^{n+1}\,(x_1 - x_2)^{-1}\,\{2\,\dot{x}_1\,\dot{x}_2$$
$$- (\omega_1)^2\,\left[x_1 + x_2 - (\lambda_1)^2\,(x_1 + x_2)^{-3}\right]x_n$$
$$+ (\omega_2)^2\,\left[x_1\,x_2 - (\lambda_2)^2\,(x_1\,x_2)^{-3}\right]\}, \quad n = 1, 2.$$

(4.110)

The explicit solution of these equations of motions reads as follows:

$$x_n(t) = -\frac{1}{2}\,\left[y_1(t) + (-1)^n\,\left\{\left[y_1(t)\right]^2 - 4\,y_2(t)\right\}^{1/2}\right], \quad n = 1, 2,$$

(4.111)

with $y_1(t)$ and $y_2(t)$ given, in terms of the *initial* data $x_1(0)$, $x_2(0)$, by the formulas (4.103) with $n = 1, 2$ where

$$y_1(0) = -[x_1(0) + x_2(0)], \quad y_2(0) = x_1(0)\, x_2(0). \qquad (4.112)$$

The necessary and sufficient condition for this solution to be *real* and *nonsingular* is validity for all time of the inequality

$$\left[y_1(t)\right]^2 > 4\, y_2(t). \qquad (4.113)$$

Let us consider firstly the *more generic* case with $(\omega_m)^2 > 0$ and $(\omega_m \lambda_m)^2 \equiv (\mu_m)^2 > 0$ for $m = 1, 2$. Then (see Remark 4.2.1) the *initial* data must satisfy the requirements $y_1(0) \neq 0$ and $y_2(0) \neq 0$ implying $x_1(0) \neq 0$, $x_2(0) \neq 0$, $x_1(0) \neq -x_2(0)$ (see (4.112)). It is then easily seen that $y_1(t) \neq 0$ and $y_2(t) \neq 0$ for all time, implying that $y_2(t) = x_1(t)\, x_2(t)$—hence also $x_1(t)$ and $x_2(t)$—do not change sign over time. Hence if $x_1(0) < 0$ and $x_2(0) > 0$ or viceversa $x_1(0) > 0$ and $x_2(0) < 0$—implying $y_2(t) < 0$—the solution (4.111) is *nonsingular*; note the *open* character of this set of *initial* data, yielding a solution in which the two particles are for *all* time on opposite sides of the origin of the *real* line and oscillate *multiperiodically*—their motions being a nonlinear superpositions of two periodic evolutions with *different* periods $T_1 = \pi / |\omega_1|$ and $T_2 = \pi / |\omega_2|$—which however, if the ratio ω_1/ω_2 is a *rational* number, is overall *periodic*, implying then that the model is *isochronous, all* its *nonsingular* solutions being *completely periodic* with a *fixed* period independent of the initial data. Note that in this case there are 4 *different equilibrium configurations,*

$$x_n(t) = \bar{x}_n = \frac{1}{2}\left\{ s_1 \left|(\lambda_1)^{1/2}\right| + s_2\, (-1)^n \left[\lambda_1 + 4\left|(\lambda_2)^{1/2}\right|\right]^{1/2} \right\}, \quad n = 1, 2, \tag{4.114a}$$

with $s_1 = \pm$ and $s_2 = \pm$ (but two of these configurations are trivially related to each other by an exchange of the labels of the two particles).

If instead initially the two particles are on the *real* line on the *same* side of the origin—i.e., $x_1(0)$ and $x_2(0)$ are *both* positive or *both* negative—then provided $(\lambda_1)^2 > 16\lambda_2$ there still is an *open* set of initial data yielding solutions which are *nonsingular*, those in which the two particles have initially *sufficiently small* velocities and are *sufficiently close* to the equilibrium configurations

$$x_n(t) = \bar{x}_n = \frac{1}{2}\left\{ s_1 \left|(\lambda_1)^{1/2}\right| \pm (-1)^n \left[\lambda_1 - 4\left|(\lambda_2)^{1/2}\right|\right]^{1/2} \right\}, \quad n = 1, 2, \tag{4.114b}$$

with $s_1 = 1$ if the two particles are (initially hence always) to the right of the origin (i.e., $x_1(t) > 0$, $x_2(t) > 0$) and $s_1 = -1$ in the opposite case ($x_1(t) < 0$, $x_2(t) < 0$).

Indeed, it is not difficult to verify that a condition on the initial data *sufficient* to guarantee that the corresponding motion be *nonsingular* is

$$(\lambda_1)^2 + 16\, h_2 > 8\, h_1 \left| (h_2)^{1/2} \right|, \qquad (4.114c)$$

with h_1 and h_2 defined in terms of the initial data by (4.103c).

Next let us consider the case with both parameters $(\omega_m)^2$ again *positive*, $(\omega_1)^2 > 0$ and $(\omega_2)^2 > 0$, but instead both parameters $\omega_m \lambda_m \equiv \mu_m$ vanishing, $\omega_1 \lambda_1 = \omega_2 \lambda_2 \equiv \mu_m = 0$. In this case the only equilibrium position is characterized by $y_m(t) = \bar{y}_m = 0$ for $m = 1, 2$, see (4.103i), and the *nonsingular* solutions are *quite nongeneric*; they clearly may only exist if ω_1 and ω_2 are *congruent* (i.e., *integer* multiples of each other), and even in such cases they require quite special sets of initial data. For instance, a simple such case would be that with $\omega_2 = 2\omega_1$; then the model would be *isochronous*, with *all* solutions *completely periodic* with period $2\pi / |\omega_2|$; but conditions on the initial data sufficient to guarantee that the solution be *nonsingular* would require the initial data to imply validity of the *equality* $\theta_2 = 2\theta_1 \bmod (2\pi)$ and of the *inequality* $(A_1)^2 > 4A_2$, see (4.103i)).

Next, let us consider the two cases with *both* parameters $(\omega_m)^2$ again *positive*, $(\omega_1)^2 > 0$ and $(\omega_2)^2 > 0$, but *only one* of the two parameters $(\omega_m \lambda_m)^2 = (\mu_m)^2$ *vanishing*. If $(\omega_1 \lambda_1)^2 \equiv (\mu_1)^2 > 0$, $\omega_2 \lambda_2 \equiv \mu_2 = 0$, the 4 equilibrium configurations of the system are $x_1 = 0$, $x_2 = \pm \left| (\lambda_1)^{1/2} \right|$ and $x_1 = \pm \left| (\lambda_1)^{1/2} \right|$, $x_2 = 0$; and the *nonsingular* solutions would be characterized by initial data $x_1(0)$, $x_2(0)$ *sufficiently close* to these equilibria and with *sufficiently small* (in modulus) initial velocities $\dot{x}_1(0)$, $\dot{x}_2(0)$, featuring coordinates $x_n(t)$ that oscillate around their equilibrium values—*multiperiodically* if ω_1 / ω_2 is *not* a *rational* number, *periodically* otherwise (*isochronous* case). If instead $\omega_1 \lambda_1 \equiv \mu_1 = 0$, $(\omega_2 \lambda_2)^2 \equiv (\mu_2)^2 > 0$, all solutions characterized by initial data featuring the two particles initially on *different* sides of the origin, so that $y_2(0) = x_1(0)x_2(0) < 0$, would be *nonsingular*, maintaining this feature throughout their time evolution—which would again be *multiperiodic* if ω_1 / ω_2 is *not* a *rational* number, *periodic* otherwise (*isochronous* case). In this case the two equilibria around which the two particles oscillate are $x_1 = s \left| (\lambda_2)^{1/4} \right|$, $x_2 = -s \left| (\lambda_2)^{1/4} \right|$ with $s = \pm$.

Next, let us consider the cases when *both* parameters ω_m vanish, $\omega_1 = \omega_2 = 0$ (excluding however the well-known case with $\omega_1 \lambda_1 = \omega_2 \lambda_2 \equiv \mu_m = 0$, corresponding to the original nonperiodic *goldfish* model). In these cases no periodic solutions exist: but, in spite of the absence of any confining potential, in this case it does *not* necessarily happen that *both* particles come in from infinitely far away in the *remote* past and escape infinitely far away in the *remote* future; although at least one must display such a *scattering* behavior. Indeed, in the case $(\omega_2 \lambda_2)^2 \equiv (\mu_2)^2 > 0$ (with $\omega_1 \lambda_1 = 0$), if the two particles are initially on opposite sides of the origin, so that $y_2(0) = x_1(0)x_2(0) < 0$, then the motions are *nonsingular* for *all* initial data; each particle remains on the same side of the origin where it was initially, one having come from infinitely far away in the *remote* past and approaching *asymptotically* the origin in the *remote* future, the other having been *asymptotically* close to the origin in the *remote* past and escaping to infinity in the *remote* future; in formulas, if $y_1(0) \neq 0$,

$$x_s(t) = -\frac{1}{2}\left\{ s \; |\dot{y}_1(0)\, t + y_1(0)| + \dot{y}_1(0)\, t + y_1(0) \right\} + O\left(\left|\frac{1}{t}\right|\right), \quad s = \pm, \quad (4.115a)$$

where we denote as $x_+(t)$ respectively as $x_-(t)$, the coordinate of the particle that goes toward *infinity* respectively toward the *origin* as $\dot{y}_1(0)\, t \to +\infty$; here $y_1(0) = -\left[x_+(0) + x_-(0)\right]$, $\dot{y}_1(0) = -\left[\dot{x}_+(0) + \dot{x}_-(0)\right]$, $y_2(0) = x_+(0)x_-(0) < 0$. While if $y_1(0) = 0$, then the behavior is *different*, with both particles coming in from *far away* in the *remote* past and returning *far away* in the *remote* future, each remaining for all time on the same side of the origin. In formulas the asymptotic behavior in this case reads as follows (as $t \to \pm\infty$):

$$\begin{aligned} x_s(t) = 2\, s\, \Big| &\left\{ (\mu_2)^2 \; [x_1(0)\, x_2(0)]^{-2} \right. \\ &\left. + [\dot{x}_1(0)\, x_2(0) + x_1(0)\, \dot{x}_2(0)]^2 \right\}^{1/2} t \Big|^{1/2} \\ &+ \frac{1}{2}\, [x_1(0) + x_2(0)] + O\left(|t|^{-1/2}\right), \quad s = \pm, \quad (4.115b) \end{aligned}$$

where now $x_+(t)$ respectively $x_-(t)$ is the coordinate of the particle on the right respectively on the left.

On the other hand if $(\omega_1\lambda_1)^2 \equiv (\mu_1)^2 > 0$ (while $\omega_2\lambda_2 \equiv \mu_2 = 0$), then as $t \to \pm\infty$:

$$\begin{aligned} x_+(t) &= [x_1(0) + x_2(0)]\, \left| \; \left|\left(1 - \rho_1^2\right)^{1/2}\right| + \frac{\mu_1\, t}{\rho_1\, [x_1(0) + x_2(0)]^2} \right| \\ &\quad + J + O\left(|t|^{-1}\right), \quad x_-(t) = -J + O\left(|t|^{-1}\right), \\ J &= \frac{[x_1(0) + x_2(0)]\; [\dot{x}_1(0)\, x_2(0) + x_1(0)\, \dot{x}_2(0)]}{\left| (\mu_1)^2 + [x_1(0) + x_2(0)]^2 \; [\dot{x}_1(0) + \dot{x}_2(0)]^2 \; \right|^{1/2}}, \quad (4.116) \end{aligned}$$

where ρ_1 is defined by (4.103j) and the label "+" denotes the particle that comes in from *far away* in the *remote* past and returns back *far away* in the *remote* future, while the other particle (identified with the label "−") always remains in a *finite* part of the real line. Note that in this case the possibility of a collision cannot be *a priori* excluded; precise conditions stating whether, for given parameters and initial data, this does or does not happen can be obtained, but they are not sufficiently enlightening to be worth reporting; it is however clear that the *nonsingular* solutions (featuring *no collision*) emerge also in this case from an *open* set of initial data.

Finally, let us consider the cases in which one of the two parameters $(\omega_m)^2$ is *positive* and the other one *vanishes*.

Firstly let us consider the cases with $(\omega_1)^2 > 0$, $\omega_2 = 0$. Let us then assume, firstly, that $(\omega_2\lambda_2)^2 \equiv (\mu_2)^2 > 0$ (while $(\omega_1\lambda_1)^2 \equiv (\mu_1)^2 \geq 0$). It is then evident that the initial data must satisfy the restriction $x_1(0) \neq 0$, $x_2(0) \neq 0$ (and, if $(\omega_1\lambda_1)^2 > 0$, also $x_1(0) \neq -x_2(0)$), and that the coordinates $x_n(t)$ cannot change sign throughout the time evolution. Then the necessary and sufficient condition for *all* solutions of the model to be *nonsingular* is that the two particles be on *different* sides of the origin. And these *nonsingular* solutions are characterized by each particle coming in from *far away* in the *remote* past and returning *far away* in the *remote* future. If instead $\omega_2\lambda_2 = \mu_2 = 0$, then conditions on the initial data *sufficient* to ensure that the solutions be *nonsingular* for $t \geq 0$ are validity of the two *inequalities*

$$x_1(0)\, x_2(0) < 0, \quad \dot{x}_1(0)\, x_2(0) + \dot{x}_2(0)\, x_1(0) < 0, \qquad (4.117)$$

with the corresponding solutions again featuring two particles flying away, one to $+\infty$ and the other to $-\infty$ as $t \to +\infty$; but these solutions would be *singular* at some *negative* time (unless the *second inequality* (4.117) became an *equality*, making the initial data *nongeneric*; the corresponding solution would then oscillate periodically in the *finite* part of the real line, with period $T_1 = \pi/|\omega_1|$ if $\lambda_1 > 0$, $2T_1$ if $\lambda_1 = 0$).

Next let us consider the cases with $\omega_1 = 0$, $(\omega_2)^2 > 0$. Then if $(\omega_2\lambda_2)^2 = (\mu_2)^2 > 0$ the origin is a forbidden point for the particles, initially and throughout the motion, hence no particle can cross the origin; therefore *all* solutions featuring the two particles on opposite sides of the origin are *nonsingular,* with one particle coming in from *far away* in the *remote* past and returning there in the *remote* future (with *asymptotically constant* velocities), and the other one approaching *asymptotically* the origin with *decreasing* velocity (and never reaching it); unless λ_1 is *finite* (so that $\omega_1\lambda_1 \equiv \mu_1 = 0$) and moreover the (*nongeneric*) initial data imply $\dot{y}_1(0) = 0$—hence $\dot{x}_1(0) = -\dot{x}_2(0)$—when the solution is *periodic* with period $T_2 = \pi/|\omega_2|$. If the particles are instead on the *same* side of the origin, then *all generic* solutions are *singular* at some time, but there is an *open* set of initial data characterized by one *inequality* (which is not sufficiently enlightening to justify explicit display)—which is sufficient to guarantee that the solutions are *nonsingular* for $t \geq 0$. The asymptotic behavior as $t \to +\infty$ of these solutions sees one coordinate (denoted below as $x_+(t)$) escaping *far away* in the *remote* future, and the other one (denoted below as $x_-(t)$) tending instead to a *finite* value. In formulas, if $(\omega_1\lambda_1)^2 = (\mu_1)^2 > 0$ then

$$x_+(t) = -\frac{\mu_1\, t}{\rho_1\, y_1(0)} + E + O\left(\frac{1}{t}\right), \quad x_-(t) = -E + O\left(\frac{1}{t}\right),$$

$$E = \frac{\rho_1\, \mu_2\, y_1(0)}{\rho_2\, \mu_1\, y_2(0)}, \quad y_1(0) \neq 0, \quad y_2(0) \neq 0; \tag{4.118a}$$

if instead $\omega_1 \lambda_1 = \mu_1 = 0$, provided $y_1(0)\dot{y}_1(0) > 0$, then

$$x_+(t) = -\left[\dot{y}_1(0)\, t + y_1(0)\right] + F + O\left(\frac{1}{t}\right), \quad x_-(t) = -F + O\left(\frac{1}{t}\right),$$

$$F = \frac{\mu_2}{\rho_2\, |y_1(0)|\, y_2(0)}, \quad y_1(0) \neq 0, \quad y_2(0) \neq 0. \tag{4.118b}$$

Here we used the above notation, see (4.103j).

The last cases to be considered are those with $\omega_1 = 0$, $(\omega_2)^2 > 0$ and $\omega_2 \lambda_2 \equiv \mu_2 = 0$. Then—if $\omega_1 \lambda_1 = \mu_1 = 0$—the solution is certainly *nonsingular* for $t \geq 0$—while it is likely to feature a *singularity* at some *negative* value of t—provided

$$[\dot{x}_1(0) + \dot{x}_2(0)]\ [x_1(0) + x_2(0)] > 0, \quad [x_1(0) + x_2(0)]^2 >$$

$$> 4\ \left| \left\{ [x_1(0)\, x_2(0)]^2 + \left[\frac{\dot{x}_1(0)\, x_2(0) + x_1(0)\, \dot{x}_2(0)}{\omega_2} \right]^2 \right\}^{1/2} \right|, \tag{4.119a}$$

featuring then, as $t \to +\infty$, one particle traveling *infinitely far away* (to the left if $x_1(0) + x_2(0) < 0$, to the right if $x_1(0) + x_2(0) > 0$) and the other one approaching *asymptotically* the origin. While if $(\omega_1 \lambda_1)^2 = (\mu_1)^2 > 0$ a condition on the initial data sufficient to guarantee that the solutions be *nonsingular* for *all* time is (in the above notation, see (4.103i, 4.103j))

$$[y_1(0)]^2\ (\rho_1)^2 > 4\, A_2; \tag{4.119b}$$

and in this case one coordinate comes in from *far away* in the *remote* past and approaches the origin in the *remote* future, while the other goes *far away* (on the opposite side) in the *remote* future and was instead *near* the origin in the *remote* past.

Let us end this discussion of the 2-body case by displaying the *Hamiltonian* that yields the equations of motion (4.110):

$$H(p_1, p_2; x_1, x_2) = \frac{1}{2} \left\{ (x_1 - x_2)^{-2} \left[\left(\frac{p_1 - p_2}{a} \right)^2 + \left(\frac{p_1\, x_1 - p_2\, x_2}{a} \right)^2 \right] \right.$$

$$+ (\omega_1)^2 \left[(x_1 + x_2)^2 + (\lambda_1)^2\, (x_1 + x_2)^{-2} \right]$$

$$\left. + (\omega_2)^2 \left[(x_1\, x_2)^2 + (\lambda_2)^2\, (x_1\, x_2)^{-2} \right] \right\}, \tag{4.120}$$

where $p_1(t)$ respectively $p_2(t)$ are the canonical momenta conjugated to the canonical coordinates $x_1(t)$ respectively $x_2(t)$ (while a is an arbitrary constant the role of which is merely to adjust the dimensions: it could be absorbed in the definition of the canonical momenta or equivalently rescaled away by renormalizing appropriately the time variable).

Let us end this Subsection 4.2.4 by underlining the complicated character of the treatment implied by our self-imposed restriction—see above—to work in the *real* context. This is an analogous phenomenon to the lack of *any simple* rule—similar to that valid in the *complex* context—to count in the *real* context the number of *zeros* of a *generic* polynomial of *arbitrary* degree N.

4.2.5 The Goldfish-CM Model

The fifth N-body problem we consider in Section 4.2 is the *integrable Hamiltonian* model characterized by the following system of *nonlinearly coupled* Newtonian equations of motion of goldfish type:

$$
\ddot{x}_n = \sum_{\ell=1,\,\ell\neq n}^{N} \left(\frac{2\,\dot{x}_n\,\dot{x}_\ell}{x_n - x_\ell} \right) - \left[\prod_{\ell=1,\,\ell\neq n}^{N} (x_n - x_\ell) \right]^{-1} \cdot
$$
$$
\cdot \sum_{m=1}^{N} \left\{ (x_n)^{N-m} \left[-\omega^2\,y_m + 2\,\lambda^2 \sum_{\ell=1,\,\ell\neq m}^{N} (y_m - y_\ell)^{-3} \right] \right\}, \quad (4.121a)
$$

with (see (2.4))

$$
y_m = (-1)^m \sum_{1 \leq s_1 < s_2 \dots < s_m \leq N} \left[\prod_{r=1}^{m} (x_{s_r}) \right]. \qquad (4.121b)
$$

Here the 2 parameters ω and λ are 2 *arbitrary real nonvanishing* numbers.

> **Remark 4.2.5.1.** These equations of motion, (4.121a), can be reformulated via (2.6) so that they read
>
> $$
> \ddot{x}_n = \sum_{\ell=1,\,\ell\neq n}^{N} \left(\frac{2\,\dot{x}_n\,\dot{x}_\ell}{x_n - x_\ell} \right) - \left[\prod_{\ell=1,\,\ell\neq n}^{N} (x_n - x_\ell) \right]^{-1} \cdot \left\{ \omega^2\,(x_n)^N \right.
> $$
> $$
> \left. + \sum_{m=1}^{N} \left[2\,\lambda^2\,(x_n)^{N-m} \sum_{\ell=1,\,\ell\neq m}^{N} (y_m - y_\ell)^{-3} \right] \right\}. \quad \blacksquare \qquad (4.121c)
> $$

The solutions $x_n(t)$ of this N-body problem are provided by the N zeros $x_n(t)$ of the (monic) polynomial (2.3) of degree N in z, the N coefficients $y_m(t)$ of which are themselves the solutions of the system of N coupled ODEs

$$\ddot{y}_m = -\omega^2 \, y_m + 2 \, \lambda^2 \sum_{\ell=1,\ \ell \neq m}^{N} (y_m - y_\ell)^{-3}, \qquad (4.122\text{a})$$

which clearly obtain from the Hamiltonian

$$H(\vec{p}, \vec{y}) = \frac{1}{2} \left\{ \sum_{m=1}^{N} \left[(p_m)^2 + (\omega \, y_m)^2 \right] + \lambda^2 \sum_{m,\ell=1,\ \ell \neq m}^{N} (y_m - y_\ell)^{-2} \right\}, \qquad (4.122\text{b})$$

where the N quantities $p_m(t)$ are the N canonical momenta conjugated to the N canonical variables $y_m(t)$.

Because this is a well-known *solvable* model (see Section 4.N), the time-dependence of these N quantities $y_m(t)$ can be obtained by solving an *algebraic* problem. Indeed the solution of the initial-value problem of this dynamical system, (4.122a), is provided by the following prescription: the N quantities $y_m(t)$ are the N *eigenvalues* of the $N \times N$ (t-dependent) matrix

$$\mathbf{Y}(t) = \mathbf{Y}(0) \, \cos(\omega \, t) + \dot{\mathbf{Y}}(0) \, \frac{\sin(\omega \, t)}{\omega}, \qquad (4.123\text{a})$$

with

$$\mathbf{Y}(0) = \text{diag} \left[y_m(0) \right], \qquad (4.123\text{b})$$

$$\dot{\mathbf{Y}}(0) = \text{diag} \left[\dot{y}_m(0) \right] + \mathbf{i} \, [\mathbf{M}(0), \, \mathbf{Y}(0)], \qquad (4.123\text{c})$$

where above and below $y_m(0)$ respectively $\dot{y}_m(0)$ are given by (2.4) respectively (2.5) in terms of the initial data $x_n(0)$ and $\dot{x}_n(0)$, while in the right-hand side of (4.123c) $[\mathbf{M}(0), \, \mathbf{Y}(0)]$ indicates the commutator of the two $N \times N$ matrices $\mathbf{M}(0)$ and $\mathbf{Y}(0)$,

$$[\mathbf{M}(0), \, \mathbf{Y}(0)] \equiv \mathbf{M}(0) \, \mathbf{Y}(0) - \mathbf{Y}(0) \, \mathbf{M}(0), \qquad (4.123\text{d})$$

with the matrix $\mathbf{M}(0)$ defined componentwise in terms of the initial data $x_n(0)$ as follows:

$$M_{nm}(0) = \lambda^2 \left[y_n(0) - y_m(0) \right]^{-2}, \quad n \neq m,$$
$$M_{nn}(0) = - \sum_{\ell=1,\ \ell \neq n} M_{n\ell}(0) = -\lambda^2 \sum_{\ell=1,\ \ell \neq n} \left[y_n(0) - y_\ell(0) \right]^{-2}, \qquad (4.123\text{e})$$

where the N numbers $y_m(0)$ are defined in terms of the N initial data $x_n(0)$ by the formulas (4.121b) at $t = 0$.

Note that these formulas, (4.123), provide an explicit definition of the N time-dependent quantities $y_m(t)$ in terms of the initial data $x_n(0)$, $\dot{x}_n(0)$ of the N-body problem of goldfish type characterized by the Newtonian equations of motion (4.121) via *algebraic* operations, amounting essentially to the solution of polynomial equations of degree N; and that the values $x_n(t)$ at time t of the particle coordinates x_n solving the *nonlinear* N-body problem of goldfish type characterized by the Newtonian equations of motion (4.121) are then provided by the N *zeros* of the polynomial (2.3), itself explicitly known in terms of its N coefficients $y_m(t)$, see (4.123).

> **Remark 4.2.5.2.** Let us call again attention, in connection with the solution— as described above—of the N-body problem of goldfish type characterized by the *Newtonian* equations of motion (4.121a), to a characteristic phenomenology also relevant to most of the models treated in Section 4.2 indeed throughout this book; but particularly relevant in the context of the model treated in Subsection 4.2.5. The identification of the N eigenvalues of a given matrix is only *unique up to permutations* (unless the corresponding eigenvector is also identified), and likewise the identification of the *zeros* of a polynomial is only *unique up to permutations*. Therefore both the N quantities $y_m(t)$ and the N coordinates $x_n = x_n(t)$ yielded by the solution as described above are only identified *up to permutations of their N labels*. But the coefficients y_m of a polynomial are instead of course an *ordered* set. Hence to solve the problem by the procedure described above it is necessary to identify the quantities $y_m(t)$ by following their evolution over time, in order to know, say, which is the coordinate $y_1(t)$ that has evolved from the coordinate $y_1(0)$; and likewise for the coordinates $x_n(t)$. On the other hand the well-known *isochronous* character of the N-body problem (4.122a)—see the relevant references in Section 4.N— implies that the N-body problem (4.121) is also *isochronous* (see Remarks 2.1.1 and 4.2). ∎

The findings reported above imply the possibility to display in *completely explicit* form the solution of the N-body problem of goldfish type characterized by the Newtonian equations of motion (4.121) for $N = 2, 3, 4$; but we doubt this would be very illuminating, and we therefore leave this task to the eager reader. We rather like to reemphasize that these findings imply that, for arbitrary N, this N-body problem is *isochronous* (provided the parameters ω and λ are both *real* and *nonvanishing*), featuring generic solutions all of which are confined to the *finite* part of the *complex x-plane* and are *completely periodic* with period $T = 2\pi / |\omega|$ (or quite possibly an *integer multiple* of this period, see Remark 4.2 and, if need be, Section 2.3).

Let us end the treatment of this model by displaying its equations of motion (4.121a) in the simplest case with $N = 2$:

$$\ddot{x}_n = \frac{(-1)^{n+1}}{x_1 - x_2} \left\{ 2\,\dot{x}_1\,\dot{x}_2 - \omega^2\,(x_n)^2 \right.$$
$$\left. + 2\,\lambda^2\,(x_1 + x_2 + x_1 x_2)^{-3}\,(1 - x_n) \right\}, \quad n = 1, 2. \qquad (4.124)$$

4.2.6 An N-Body Problem Involving N^2 Arbitrary Coupling Constants

The sixth example we report in this Section 4.2 is a model featuring N^2 *arbitrary* coupling constants a_{mj}. Its equations of motion read

$$\ddot{x}_n = \sum_{\ell=1,\,\ell\neq n}^{N} \left(\frac{2\,\dot{x}_n\,\dot{x}_\ell}{x_n - x_\ell} \right) - \left[\prod_{\ell=1,\,\ell\neq n}^{N} (x_n - x_\ell) \right]^{-1} \sum_{m,j=1}^{N} \left[a_{mj}\, y_j\, (x_n)^{N-m} \right], \quad (4.125)$$

with the N quantities $y_j(t)$ replaced by their expressions in terms of the N particle coordinates $x_\ell(t)$, see (2.4). The corresponding equations of motion for the quantities $y_m(t)$ read

$$\ddot{y}_m = \sum_{j=1}^{N} \left(a_{mj}\, y_j \right), \quad (4.126a)$$

implying that their solutions read

$$y_m(t) = \sum_{n=1}^{N} \left\{ \left[y_n^{(+)} \exp\left(\mathbf{i}\,\alpha_n\, t \right) + y_n^{(-)} \exp\left(-\mathbf{i}\,\alpha_n\, t \right) \right] u_m^{(n)} \right\}. \quad (4.126b)$$

Here the quantities α_n respectively $u_m^{(n)}$ are defined via the N eigenvalues $(\alpha_n)^2$ respectively the corresponding N eigenvectors $\mathbf{u}^{(n)}$ (with components $u_m^{(n)}$) of the (time-independent) eigenvalue problem

$$\mathbf{A}\,\mathbf{u}^{(n)} = \alpha_n^2\,\mathbf{u}^{(n)}, \quad (4.127)$$

where the $N \times N$ matrix $\mathbf{A}$ features as its N^2 elements the N^2 coupling constants a_{nm}, see (4.126a); while the $2N$ time-independent quantities $y_n^{(\pm)}$ are the solutions of the following linear system of $2N$ algebraic equations (see (4.126b)):

$$\sum_{n=1}^{N} \left\{ \left[y_n^{(+)} + y_n^{(-)} \right] u_m^{(n)} \right\} = y_m(0), \quad (4.128a)$$

$$\mathbf{i} \sum_{n=1}^{N} \left\{ \alpha_n \left[y_n^{(+)} - y_n^{(-)} \right] u_m^{(n)} \right\} = \dot{y}_m(0), \quad (4.128b)$$

where the $2N$ quantities $y_m(0)$ respectively $\dot{y}_m(0)$ are expressed in terms of the initial data $x_n(0)$, $\dot{x}_m(0)$ via (2.4) and (2.5).

Note that in writing this solution, (4.126b), we are implicitly assuming that the N eigenvalues $(\alpha_n)^2$ are *all different among themselves*; otherwise the standard limit must be taken, yielding terms in the right-hand side of (4.126b) featuring powers of t.

Remark 4.2.6.1. Clearly a rescaling respectively a shifting of the N coordinates $x_n(t)$,

$$x_n(t) \Rightarrow \beta_n \, x_n(t) + \gamma_n \tag{4.129}$$

with β_n respectively γ_n arbitrary (constant) parameters, changes only rather trivially the equations of motions (4.126a): indeed if $\beta_n = \beta$ and $\gamma_n = \gamma$, it amounts almost only to a redefinition of the N^2 coupling constants a_{nm}. ∎

Remark 4.2.6.2. The above technique of solution provides the *configuration* of the system at time t as an *unordered* set of N coordinates $x_n(t)$, but it does *not* allow to identify, say, which is the value at time t of the specific coordinate $x_1(t)$ which has evolved by *continuity* in time from the initial data $x_1(0)$, $\dot{x}_1(0)$: see Remark 2.1.1, and the following Remark 4.2.6.3. ∎

Remark 4.2.6.3. The above treatment shows that the following *general properties* of the N-body problem characterized by the Newtonian equations of motion (4.125) hold.

(i) For *generic (complex)* initial data $x_n(0)$, $\dot{x}_n(0)$ the time evolution of this N-body system is, for *all* (finite) time, *nonsingular*.

(ii) If the N eigenvalues $(\alpha_n)^2$ of the $N \times N$ matrix **A** are *all positive* and *distinct*— $(\alpha_n)^2 > 0$, $(\alpha_n)^2 \neq (\alpha_m)^2$ if $n \neq m$—then for *generic (complex)* initial data the time evolution of this N-body system is *multiply periodic*, hence *confined* to a *finite* region of its phase space.

(iii) If and only if some (say, ν) of the N eigenvalues α_n^2 of the $N \times N$ matrix **A** are— up to a *common positive* factor α^2—the *squares of distinct rational numbers*, then the system (4.126a) features ν *completely periodic* solutions.

(iv) If and only if *all* the N eigenvalues $(\alpha_n)^2$ of the $N \times N$ matrix **A** are—up to a *common positive factor* α^2—the *squares of N distinct rational numbers*— i.e., $(\alpha_n)^2 = (\rho_n \alpha)^2$ with all the parameters ρ_n *nonvanishing rational* numbers and α an *arbitrary nonvanishing real number*—then the system (4.126a) is *isochronous*: *all* its *generic* solutions are *nonsingular* and *completely periodic* with a period—an *integer multiple* of $2\pi / |\alpha|$—independent of the initial data. ∎

The enterprising reader is invited to generalize this Remark 4.2.6.3 to the case in which the quantities $(\alpha_n)^2$ are *not all real*, as might well be the case if the matrix **A** is *real* but *not symmetrical*.

4.3 Solvable *N*-Body Problems with Additional Velocity-Dependent Forces

In Section 4.3 we treat a few *solvable N*-body problems characterized by the presence of additional *velocity-dependent* terms besides the standard one identifying these models as being *of goldfish type*. In particular in Section 4.3 we take as starting points of our treatments some *solvable* examples belonging to the following class of *decoupled second-order* ODEs satisfied by the coefficients $y_m(t)$ of the polynomial (2.3):

$$\ddot{y}_m = f_m(y_m, \dot{y}_m); \qquad\qquad (4.130)$$

and—when appropriate—to the corresponding *isochronous* versions of these ODEs.

4.3.1 A Very Simple Example

The *first* assignment in (4.130) reads

$$f_m(y, \dot{y}) = (\mathbf{i}\,\omega_m - \lambda_m)\,\dot{y} + \mathbf{i}\,\omega_m\lambda_m\,y, \qquad\qquad (4.131a)$$

so that

$$\ddot{y}_m = (\mathbf{i}\,\omega_m - \lambda_m)\,\dot{y} + \mathbf{i}\,\omega_m\lambda_m\,y, \qquad\qquad (4.131b)$$

with the $2N$ parameters ω_m and λ_m *a priori* arbitrarily assigned (but see below). Then the *Newtonian* equations of motion of the N-body problems characterizing the motion of the N zeros $x_n(t)$ of the polynomial (2.3) read (see (4.79))

$$\ddot{x}_n = \sum_{\ell=1,\,\ell\neq n}^{N} \left(\frac{2\,\dot{x}_n\,\dot{x}_\ell}{x_n - x_\ell} \right) - \left[\prod_{\ell=1,\,\ell\neq n}^{N} (x_n - x_\ell) \right]^{-1} \cdot$$
$$\cdot \sum_{m=1}^{N} \left\{ \left[(\mathbf{i}\,\omega_m - \lambda_m)\,\dot{y}_m + \mathbf{i}\,\omega_m\,\lambda_m\,y_m \right] (x_n)^{N-m} \right\}, \qquad (4.131c)$$

with the N coefficients $y_m(t)$ expressed via the N coordinates $x_n(t)$ by (2.4) and their N time-derivatives $\dot{y}_m(t)$ expressed via the N coordinates $x_n(t)$ and their N time derivatives $\dot{x}_n(t)$ via (2.5).

The solution of this Newtonian N-body model is therefore provided by the N zeros $x_n(t)$ of the polynomial (2.3) of degree N in z with

$$y_m(t) = y_m(0)\,\exp(-\lambda_m\,t) + \left[\frac{\lambda_m\,y_m(0) + \dot{y}_m(0)}{\lambda_m + \mathbf{i}\,\omega_m} \right] \cdot$$
$$\cdot \left[\cos(\omega_m\,t) + \mathbf{i}\,\sin(\omega_m\,t) - \exp(-\lambda_m\,t) \right], \qquad (4.131d)$$

where the quantities $y_m(0)$ are expressed in terms of the initial values $x_n(0)$ of the coordinates $x_n(t)$ by (2.4) and the quantities $\dot{y}_m(0)$ are expressed in terms of the initial values $x_n(0)$ of the coordinates $x_n(t)$ and the initial values $\dot{x}_n(0)$ of the velocities $\dot{x}_n(t)$ by (2.5). Indeed—as the diligent reader will readily verify—these formulas provide the solution of the initial-value problem of the system of ODEs (4.130) with (4.131a).

Note that this implies that *all* solutions of the *N*-body problem characterized by these Newtonian equations of motion, as seen in (4.131c), are *multiply periodic*, respectively *asymptotically multiply periodic*, if the 2*N* parameters ω_m and λ_m are *all real* and satisfy the requirements $\omega_m \neq 0$ and $\lambda_m = 0$ respectively $\lambda_m > 0$; while the model is *isochronous*, respectively *asymptotically isochronous,* if the *N* parameters ω_m satisfy the Diophantine restriction $\omega_m = r_m \omega$ with the numbers r_m *all nonvanishing rational* numbers, ω an *arbitrary real nonvanishing* constant, and $\lambda_m = 0$ respectively $\lambda_m > 0$ (for a precise definition of *asymptotic isochrony* and other examples of models featuring this property, see [54] and Chapter 6 of [27]).

4.3.2 Another Simple Example and Its Isochronous Variant

This model obtains from the following *second* assignment in (4.130):

$$f_m (y, \dot{y}) = c_m (\dot{y})^2 y^{-1}, \tag{4.132a}$$

featuring the *N*, *a priori arbitrary,* coupling constants c_m. This implies that the ODEs determining the time evolution of the quantities $y_m(t)$ read

$$\ddot{y}_m = c_m (\dot{y}_m)^2 (y_m)^{-1}, \tag{4.132b}$$

so that their solutions—see Subsection 2.4.1—read

$$y_m(t) = y_m(0) \left(1 - \frac{t}{\bar{t}_m}\right)^{1/(1-c_m)}, \quad \bar{t}_m = \frac{y_m(0)}{(c_m - 1) \dot{y}_m(0)}. \tag{4.132c}$$

Here and hereafter we assume for simplicity that none of the *N* parameters c_m vanishes nor equals unity, $c_m \neq 0$, $c_m \neq 1$.

The corresponding *solvable N*-body problem of goldfish type satisfied by the *N* particle coordinates $x_n(t)$ then reads as follows:

$$\ddot{x}_n = \sum_{\ell=1, \ \ell \neq n}^{N} \left(\frac{2 \dot{x}_n \dot{x}_\ell}{x_n - x_\ell}\right) - \left[\prod_{\ell=1, \ \ell \neq n}^{N} (x_n - x_\ell)\right]^{-1} \cdot$$
$$\cdot \sum_{m=1}^{N} \left[c_m (\dot{y}_m)^2 (y_m)^{-1} (x_n)^{N-m}\right], \tag{4.133a}$$

with the quantities $y_m(t)$ respectively $\dot{y}_m(t)$ expressed in terms of the coordinates $x_n(t)$ and their time derivatives by the *explicit* formulas (2.4) respectively (2.5); and its solution is provided by the *N zeros* $x_n(t)$ of the polynomial (2.3) with the coefficients $y_m(t)$ given by the explicit formulas (4.132c) where the quantities $y_m(0)$

respectively $\dot{y}_m(0)$ are now expressed in terms of the coordinates $x_n(0)$ and $\dot{x}_n(0)$ by the *explicit* formulas (2.4) respectively (2.5).

Let us display the explicit form of these equations of motion in the simplest $N = 2$ case:

$$\ddot{x}_n = \left[\frac{(-1)^{n+1}}{x_1 - x_2} \right] \left[2\,\dot{x}_1\,\dot{x}_2 + \frac{c_1\,(\dot{x}_1 + \dot{x}_2)^2\,x_n}{x_1 + x_2} - \frac{c_2\,(\dot{x}_1\,x_2 + \dot{x}_2\,x_1)^2}{x_1\,x_2} \right],$$
$$n = 1, 2\,.$$

$$(4.133b)$$

A more general—perhaps more interesting—model is obtained by replacing the ODEs (4.132b) with the *isochronous* ODEs

$$\ddot{y}_m = \mathbf{i}\,[2\,r_m\,(1 - c_m) + 1]\,\rho_m\,\omega\,\dot{y}_m + r_m\,[r_m\,(1 - c_m) + 1]\,(\rho_m\,\omega)^2\,y_m$$
$$+ c_m\,(\dot{y}_m)^2\,(y_m)^{-1}$$

$$(4.134)$$

(see again Subsection 2.4.1, and note the replacement there of ω with $\omega_m = \rho_m\,\omega$, allowing thereby the introduction of the N additional parameters ρ_m). In this ODEs (4.134) ω is an *arbitrary nonvanishing real* parameter, to which we associate as usual the period

$$T = \frac{2\pi}{|\omega|},$$

$$(4.135)$$

and the $2N$ parameters r_m and ρ_m are $2N$ *arbitrary nonvanishing rational* numbers,

$$r_m = \frac{q_m}{k_m}, \quad \rho_m = \frac{\lambda_m}{\mu_m},$$

$$(4.136)$$

where q_m and k_m are two *coprime integers* ($k_m \geq 1$, $q_m \neq 0$) and likewise λ_m and μ_m are two *coprime integers* ($\mu_m \geq 1$, $\lambda_m \neq 0$).

The solutions of these ODEs (4.134) read as follows (see, if need be, Subsection 2.4.1, taking account of notational changes):

$$y_m(t) = \exp\left(\mathbf{i}\,r_m\,\omega\,t\right)\,\gamma_m\left(\tau_m(t)\right), \quad \tau_m(t) = \frac{\exp\left(\mathbf{i}\,\rho_m\,\omega\,t\right) - 1}{\mathbf{i}\,\rho_m\,\omega},$$

$$(4.137a)$$

$$\gamma_m\left(\tau\right) = y_m(0)\,\left(1 - \frac{\tau}{\bar{\tau}_m}\right)^{1/(1-c_m)}$$
$$= y_m(0)\,\left(\mathbf{i}\,\rho_m\,\omega\,\bar{\tau}_m\right)^{1/(c_m - 1)}\,\left[\bar{\eta}_m - \exp\left(\mathbf{i}\,\rho_m\,\omega\,t\right)\right]^{1/(1-c_m)},$$

$$(4.137b)$$

$$\bar{\tau}_m = \frac{y_m(0)}{(c_m - 1)\,\left[\dot{y}_m(0) - \mathbf{i}\,r_m\,\rho_m\,\omega\,y_m(0)\right]}, \quad \bar{\eta}_m = 1 + \mathbf{i}\,\rho_m\,\omega\,\bar{\tau}_m\,.$$

$$(4.137c)$$

It is now clear via the key formulas (2.8) that the corresponding *solvable N*-body problem for the *N* particle coordinates $x_n(t)$ reads as follows:

$$
\ddot{x}_n = \sum_{\ell=1,\,\ell\neq n}^{N} \left(\frac{2\,\dot{x}_n\,\dot{x}_\ell}{x_n - x_\ell}\right) - \left[\prod_{\ell=1,\,\ell\neq n}^{N}(x_n - x_\ell)\right]^{-1} \cdot
$$

$$
\cdot \sum_{m=1}^{N} \{\mathbf{i}\,[2\,r_m\,(1 - c_m) + 1]\,\rho_m\,\omega\,\dot{y}_m
$$

$$
+ r_m\,[r_m\,(1 - c_m) + 1]\,(\rho_m\,\omega)^2\,y_m
$$

$$
+ c_m\,(\dot{y}_m)^2\,(y_m)^{-1}\}\,(x_n)^{N-m}, \tag{4.138}
$$

where $y_m(t)$ respectively $\dot{y}_m(t)$ must be replaced by their expressions (2.4) respectively (2.5).

Remark 4.3.2.1.   It is clear from (4.137) that, if the initial data $y_m(0)$ and $\dot{y}_m(0)$ imply

$$
|\bar{\eta}_m| > 1 \tag{4.139}
$$

(see (4.137c)), then $y_m(t)$ is *periodic* in time with at most (see Remark 2.4.1.2) *period* $M_m T$ where

$$
M_m = \text{MinimumCommonMultiple}\,[k_m,\ \mu_m] \tag{4.140}
$$

(see (4.137a) and the second equality (4.137b) with (4.136) and (4.135)). Hence, if for all values of *m* in its range from 1 to *N* there holds the inequality (4.139)—and note that this restriction is satisfied for an *open* set of *initial* data $y_m(0)$ and $\dot{y}_m(0)$, corresponding to an *open* set of *initial* data $x_n(0)$ and $\dot{x}_n(0)$, see (2.4) and (2.5)— then clearly the polynomial (2.3) with coefficients $y_m(t)$ is periodic with at most (see Remark 2.4.1.2) period *KT* where *K* is the Minimum Common Multiple of the *N* integers M_m, see (4.140). Hence for this *open* set of initial data $x_n(0)$ and $\dot{x}_n(0)$, the solution of the *N*-body problem of goldfish type characterized by the Newtonian equations of motion (4.138)—given by the *N* zeros $x_n(t)$ of the polynomial (2.3) with its *N* coefficients $y_m(t)$ given by (4.137)—is *periodic*, with at most (see Remark 2.4.1.2) a period νKT—where ν an *integer* in the range $1 \leq \nu \leq \nu_{\text{Max}}(N)$, see Section 2.3—which is independent from the values of the initial data, provided they belong to the appropriate *open* set defined above. This fact qualifies this *N*-body problem as *isochronous*.

This model behaves *isochronously* for an *arbitrary* set of initial data provided the *N*, a priori *arbitrary*, parameters c_m are also *restricted*—as the 2*N* parameters r_m and ρ_m (see (4.136))—to be *real rational numbers*; in which case clearly *all* the *nonsingular* solutions of this model are *completely periodic* with a period which is an *integer multiple* of the basic period *T*, see (4.135); while the *singular* solutions emerge from a set of *initial* data having *positive codimension* in the space of such data. See Subsection 2.4.1 for a more complete discussion of this case, and also of the *asymptotically isochronous* variant of this model which prevails if the parameter ω is *not real*. ∎

We complete this Subsection 4.3.2 by displaying the equations of motion (4.138) in the simplest case with $N = 2$:

$$
\begin{aligned}
\ddot{x}_n = \frac{(-1)^{n+1}}{x_1 - x_2} \Big[& 2 \ (1 - c_2) \ \dot{x}_1 \ \dot{x}_2 + \big\{ \mathbf{i} \ [2 \ r_1 \ (1 - c_1) + 1] \ \rho_1 \ \omega \ (\dot{x}_1 + \dot{x}_2) \\
& + r_1 \ [r_1 \ (1 - c_1) + 1] \ (\rho_1 \ \omega)^2 \ (x_1 + x_2) + c_1 \ (\dot{x}_1 + \dot{x}_2)^2 \ (x_1 + x_2)^{-1} \big\} \ x_n \\
& - \big\{ \mathbf{i} \ [2 \ r_2 \ (1 - c_1) + 1] \ \rho_2 \ \omega \ (\dot{x}_1 \ x_2 + \dot{x}_2 \ x_1) \\
& + r_2 \ [r_2 \ (1 - c_2) + 1] \ (\rho_2 \ \omega)^2 \ x_1 \ x_2 \\
& + c_2 \ \big[(\dot{x}_1)^2 \ (x_1)^{-1} \ x_2 + (\dot{x}_2)^2 \ (x_2)^{-1} \ x_1 \big] \big\} \Big].
\end{aligned}
\tag{4.141}
$$

4.3.3 Yet Another Simple Isochronous Model

For the *third* model treated in Section 4.3, we take as starting point the following set of ODEs:

$$
\ddot{y}_m = a_m \ \dot{y}_m \ \dot{y}_j \ \left(y_j \right)^{-1},
\tag{4.142a}
$$

where j is a *fixed* index in the range from 1 to N. We moreover assume again, for simplicity, that *all* the N parameters a_m do *not* vanish, $a_m \neq 0$, and moreover that the parameter a_j equals unity, $a_j = 1$, leaving to the interested reader the investigation of the more general case (if need be, with the help of the treatment of Subsection 2.4.1, the main results of which are tersely reported here, with the appropriate notational changes).

Hence for $m = j$ this ODE, (4.142a), reads as follows:

$$
\ddot{y}_j = \left(\dot{y}_j \right)^2 \ \left(y_j \right)^{-1},
\tag{4.142b}
$$

and it is easily seen that its solution reads

$$
y_j(t) = y_j(0) \ \exp \left[\frac{\dot{y}_j(0) \ t}{y_j(0)} \right],
\tag{4.142c}
$$

while the solutions of the remaining $N - 1$ ODEs (4.142a)—for $m \neq j$—read

$$
y_m(t) = y_m(0) + \dot{y}_m(0) \ \left[\frac{\exp \left(A_{jm} \ t \right) - 1}{A_{jm}} \right], \quad A_{jm} = \frac{a_m \ \dot{y}_j(0)}{y_j(0)}, \quad m \neq j.
\tag{4.142d}
$$

We now proceed immediately to the *isochronous* variant, which obtains in the standard manner (see Section 2.4), i.e., via the *formal* replacement of t with $\tau(t)$ and of $y_m(t)$ with $\gamma_m(\tau)$, followed by the positions

$$y_m(t) = \exp\left(\mathbf{i}\, r_m\, \omega\, t\right)\, \gamma_m\left(\tau\right), \quad \tau = \frac{\exp\left(\mathbf{i}\, \omega\, t\right) - 1}{\mathbf{i}\, \omega}, \tag{4.143a}$$

with ω an *arbitrary real nonvanishing* parameter to which we associate as usual the period $T = 2\pi/|\omega|$, and with the N parameters r_m *real* and *rational*, but otherwise *arbitrarily assigned,*

$$r_m = \frac{q_m}{k_m}, \tag{4.143b}$$

with q_m and k_m two *coprime integers* ($k_m \geq 1$, $q_m \neq 0$). This entails the replacement of the ODEs (4.142a) with

$$\ddot{y}_m = \left(2\, r_m + 1 - a_m\, r_j\right) \mathbf{i}\, \omega\, \dot{y}_m + r_m\, \left(r_m + 1 - a_m\, r_j\right)\, \omega^2\, y_m$$
$$+ a_m\, \left(\dot{y}_m - \mathbf{i}\, r_m\, \omega\, y_m\right)\, \dot{y}_j\, \left(y_j\right)^{-1}, \tag{4.144a}$$

and the following expression for the solutions of this system of ODEs

$$y_j(t) = y_j(0)\, \exp\left(\mathbf{i}\, r_j\, \omega\, \tau(t)\right)\, \exp\left\{\left[\frac{\left[\dot{y}_j(0) - \mathbf{i}\, r_j\, \omega\, y_j(0)\right]\, \tau(t)}{y_j(0)}\right]\right\}, \tag{4.144b}$$

$$y_m(t) = y_m(0) + \left[\dot{y}_m(0) - \mathbf{i}\, r_m\, \omega\, y_m(0)\right]\, \left\{\frac{\exp\left[A_{jm}\, \tau(t)\right] - 1}{A_{jm}}\right\},$$

$$A_{jm} = \frac{a_m\, \left[\dot{y}_j(0) - \mathbf{i}\, r_j\, \omega\, y_j(0)\right]}{y_j(0)}, \quad m \neq j, \tag{4.144c}$$

where

$$\tau(t) = \frac{\exp\left(\mathbf{i}\, \omega\, t\right) - 1}{\mathbf{i}\, \omega}. \tag{4.144d}$$

These formulas show that the time-dependent monic polynomial (2.3) featuring these N *coefficients* $y_m(t)$ is *nonsingular* and *completely periodic* with at most (see Remark 2.4.1.2) period KT, where K is the Minimum Common Multiple of the N integers k_m, see (4.143b).

The corresponding N-body problem of goldfish type satisfied by the N zeros $x_n(t)$ of this polynomial (2.3) reads, via (2.8), as follows:

$$\ddot{x}_n = \sum_{\ell=1,\ \ell\neq n}^{N} \left(\frac{2\, \dot{x}_n\, \dot{x}_\ell}{x_n - x_\ell}\right) - \left[\prod_{\ell=1,\ \ell\neq n}^{N} \left(x_n - x_\ell\right)\right]^{-1}$$

$$\times \sum_{m=1}^{N} \left\{\left[\mathbf{i}\, \left(2\, r_m + 1 - a_m\, r_j\right)\, \omega\, \dot{y}_m + r_m\, \left(r_m + 1 - a_m\, r_j\right)\, \omega^2\, y_m\right.\right.$$

$$\left.\left. + a_m\, \left(\dot{y}_m - \mathbf{i}\, r_m\, \omega\, y_m\right)\, \dot{y}_j\, \left(y_j\right)^{-1}\right]\, \left(x_n\right)^{N-m}\right\}, \tag{4.145}$$

where $a_j = 1$ while the other $N - 1$ parameters a_m (with $m \neq j$) are *arbitrary nonvanishing complex* numbers, and the N parameters r_m are N *arbitrary real rational* numbers, see (4.143b). It is as usual understood that the quantities $y_m(t)$ respectively $\dot{y}_m(t)$ appearing in these formulas are replaced by their expressions (2.4) and (2.5).

This N-body problem is clearly *isochronous*; we leave to the interested reader a more detailed discussion of its behavior (see Section 2.3).

We complete this Subsection 4.3.3 by displaying these equations of motion (4.145) in the simple case with $N = j = 2$:

$$
\begin{aligned}
\ddot{x}_n = \frac{(-1)^{n+1}}{x_1 - x_2} \Big[& \big\{ \mathbf{i}\,(2\,r_1 + 1 - a_1\,r_2)\,\omega\,(\dot{x}_1 + \dot{x}_2) \\
& + r_1\,(r_1 + 1 - a_1\,r_2)\,\omega^2\,(x_1 + x_2) \\
& + a_1\,[\dot{x}_1 + \dot{x}_2 - \mathbf{i}\,r_1\,\omega\,(x_1 + x_2)]\,\big[\dot{x}_1\,(x_1)^{-1} + \dot{x}_2\,(x_2)^{-1}\big] \big\}\,x_n \\
& - \mathbf{i}\,\rho_2\,\omega\,(\dot{x}_1\,x_2 + x_1\,\dot{x}_2) - r_2\,\omega^2\,x_1\,x_2 \\
& - (\dot{x}_1)^2\,(x_1)^{-1}\,x_2 - (\dot{x}_2)^2\,(x_2)^{-1}\,x_1 \Big], \quad n = 1, 2.
\end{aligned}
\tag{4.146}
$$

4.3.4 A Solvable N-Body Problem of Goldfish Type Featuring Many Arbitrary Parameters, and Its Isochronous Variant

The *fourth* model of Section 4.3 obtains from the following assignment in (4.130):

$$
f_m\,(y;\,\dot{y}) = c_m\,(\dot{y})^{\alpha_m}, \tag{4.147a}
$$

implying

$$
\ddot{y}_m = c_m\,(\dot{y})^{\alpha_m}. \tag{4.147b}
$$

Here the $2N$ parameters c_m and α_m are *arbitrary* ($c_m \neq 0, \alpha_m \neq 1, 2$; for background material relevant to the treatment of this Subsection 4.3.4 see Subsection 2.4.3 and, for the special case $\alpha_m = 2$, Subsection 2.4.2). Then the Newtonian equations of motion of the N-body problem characterizing the motion of the N zeros $x_n(t)$ of the monic polynomial (2.3), the coefficients $y_m(t)$ of which evolve according to the ODEs (4.147b), read (see (4.79))

$$
\ddot{x}_n = \sum_{\ell = 1,\,\ell \neq n}^{N} \left(\frac{2\,\dot{x}_n\,\dot{x}_\ell}{x_n - x_\ell} \right) - \left[\prod_{\ell = 1,\,\ell \neq n}^{N} (x_n - x_\ell) \right]^{-1} \sum_{m=1}^{N} \left[c_m\,(\dot{y}_m)^{\alpha_m}\,(x_n)^{N-m} \right],
\tag{4.147c}
$$

with $\dot{y}_m$ expressed via the N coordinates x_n and their N time-derivatives $\dot{x}_n$ via (2.5).

The solution of this Newtonian N-body system of goldfish type is therefore provided by the N zeros $x_n(t)$ of the polynomial of degree N in z (see (2.3)) with the N coefficients $y_m(t)$—solution of the ODE (4.147b), the initial-value problem of which is *solvable* (see Subsection 2.4.3)—reading as follows:

$$y_m(t) = y_m(0) + \dot{y}_m(0) \left(\frac{\alpha_m - 1}{\alpha_m - 2} \right) \bar{t}_m \left[1 - \left(1 - \frac{t}{\bar{t}_m} \right)^{(\alpha_m - 2)/(\alpha_m - 1)} \right],$$

$$\bar{t}_m = \frac{[\dot{y}_m(0)]^{1 - \alpha_m}}{c_m\,(\alpha_m - 1)}. \tag{4.147d}$$

To get the solution of the *initial-value* problem for the system of Newtonian equations (4.147c), in these formulas (4.147d) the initial values $y_m(0)$ and $\dot{y}_m(0)$ of the coefficients $y_m(t)$ must be expressed in terms of the initial values $x_n(0)$ and $\dot{x}_n(0)$ via the standard equations (2.4) and (2.5) at $t = 0$.

An *isochronous* variant of this model can be obtained via the same trick already used in Subsections 4.2.3 and 4.3.2 and described in Section 2.4, entailing the convenient replacement of the parameters α_m with the parameters r_m related to α_m as follows:

$$r_m = \frac{\alpha_m - 2}{1 - \alpha_m}, \quad \alpha_m = \frac{r_m + 2}{r_m + 1}, \tag{4.148}$$

together with the usual assumption that all the parameters r_m be *real* and *rational* (and $r_m \neq 0$; see below for the $r_m = 0$ case implying $\alpha_m = 2$). The relevant results—that the diligent reader will easily verify, if need be with the help of Subsection 2.4.3 (with the replacement there of ω with $\omega_m = \rho_m \omega$, allowing thereby the introduction of the N additional parameters ρ_m, see below)—are reported here via the following

Proposition 4.3.4.1. For *generic* initial data $x_n(0)$ and $\dot{x}_n(0)$, the solution of the N-body problem characterized by the following Newtonian equations of motions,

$$\ddot{x}_n = \sum_{\ell=1,\ \ell \neq n}^{N} \frac{2\, \dot{x}_n\, \dot{x}_\ell}{x_n - x_\ell} - \left[\prod_{\ell=1,\ \ell \neq n}^{N} (x_n - x_\ell) \right]^{-1} \cdot$$

$$\cdot \sum_{m=1}^{N} \left\{ \left[\mathbf{i}\,(2\,r_m + 1)\,\rho_m\,\omega\,\dot{y}_m + r_m\,(r_m + 1)\,(\rho_m\,\omega)^2\,y_m \right. \right.$$

$$\left. \left. + c_m\,(\dot{y}_m - \mathbf{i}\,r_m\,\rho_m\,\omega\,y_m)^{(2+r_m)/(1+r_m)} \right] (x_n)^{N-m} \right\} \tag{4.149a}$$

is provided by the N zeros $x_n(t)$ of the polynomial (2.3) the N *coefficients* $y_m(t)$ of which are given by the formulas

$$y_m(t) = \tilde{y}_m(\eta_m), \quad \eta_m \equiv \eta_m(t) = \exp(\mathbf{i}\,\rho_m\,\omega\,t), \tag{4.149b}$$

$$\tilde{y}_m(\eta) = \eta^{r_m}\left\{ y_m(0) + \left(\frac{1+r_m}{c_m}\right)\left[\dot{y}_m(0) - \mathbf{i}\,r_m\,\rho_m\,\omega\,y_m(0)\right]^{-r_m/(1+r_m)} \right.$$

$$\left. \cdot \left[\left(\frac{\bar{\eta}_m - \eta}{\bar{\eta}_m - 1}\right)^{-r_m} - 1\right]\right\}, \tag{4.149c}$$

$$\bar{\eta}_m = 1 + \mathbf{i}\,\omega\left(\frac{1+r_m}{c_m}\right)\left[\dot{y}_m(0) - \mathbf{i}\,r_m\,\rho_m\,\omega\,y_m(0)\right]^{-1/(1+r_m)}. \tag{4.149d}$$

In these formulas ω is a *real nonvanishing* parameter to which the basic period $T = 2\pi/|\omega|$ is associated, and the $2N$ parameters r_m and ρ_m are *arbitrary real rational* numbers,

$$r_m = \frac{q_m}{k_m}, \quad \rho_m = \frac{\lambda_m}{\mu_m}, \tag{4.149e}$$

with q_m and k_m two *coprime integers* ($k_m \geq 1$, $q_m \neq 0$) and likewise λ_m and μ_m two *coprime integers* ($\mu_m \geq 1$, $\lambda_m \neq 0$). In all these formulas (4.149) the quantities $y_m(t)$ respectively $\dot{y}_m(t)$ are supposed to be replaced by their expressions (2.4) respectively (2.5) in terms of the particle coordinates $x_n(t)$ and $\dot{x}_n(t)$ (at the time t in (4.149a), at the initial time $t = 0$ in (4.149c) and (4.71)). ∎

Remark 4.3.4.1. For the following 4 values of the parameter, $r_m = 0, -2/3, -4/3, -2$, the corresponding exponent $\lambda_m = (2 + r_m)/(1 + r_m)$ appearing in the right-hand side of the equations of motion (4.149a) takes an *integer* value: $\lambda_m = 2, 4, -2, 0$. ∎

The *isochronous* character of this model is evident from this solution. Indeed in this case the monic polynomial (2.3) with the N coefficients $y_m(t)$, see (4.149b), is *periodic* with at most (see Remark 2.4.1.2) period MT, where T is the basic period, $T = 2\pi/|\omega|$, and M is the Minimum Common Multiple of the $2N$ integers k_m and μ_m, see (4.149e)—excluding only a *nongeneric* set of initial data yielding *singularities* of the equations of motion or of the solutions. Hence the solutions of the N-body model (4.149) are *periodic* with at most (see Remark 2.4.1.2) *period* $\nu M T$, where ν is an *integer* in the range $1 \leq \nu \leq \nu_{\text{Max}}(N)$ (the value of which depends—but generally not very sensitively—on the assignment of the initial data: see Section 2.3). This implies that *all* the solutions of the N-body model (4.149)—except for a *nongeneric* set of *singular* solutions—are *periodic* with, at most (see Remark 2.4.1.2), the *period* $T_{\text{Max}} = \nu_{\text{Max}}(N)\,M\,T$, $x_n(t + T_{\text{Max}}) = x_n(t)$: *isochrony!*

Let us complete the treatment of this model by including the possibility that some of the parameters r_m vanish, $r_m = 0$, implying that correspondingly $\alpha_m = 2$ (see (4.148)). Then—see Subsection 2.4.2—the corresponding expressions (4.149b) of the functions $y_m(t)$ are replaced as follows:

$$y_m(t) = y_m(0) - (c_m)^{-1}\ln\left[\frac{\hat{\eta}_m - \exp(\mathbf{i}\,\rho_m\,\omega\,t)}{\hat{\eta}_m - 1}\right], \quad \hat{\eta}_m = 1 + \frac{\mathbf{i}\,\rho_m\,\omega}{c_m\,\dot{y}_m(0)}. \tag{4.150}$$

Hence, as discussed in Subsection 2.4.2, this solution $y_m(t)$ is *nonsingular*—except for the *nongeneric* set of initial data $\dot{y}_m(0)$ such that $\left|\hat{\eta}_m\right| = 1$—and, if the initial datum $\dot{y}_m(0)$ implies the *inequality* $\left|\hat{\eta}_m\right| > 1$, it is again *periodic* with at most (see Remark 2.4.1.2) period $\mu_m\,T$ where $T = 2\pi/\omega$ is the basic period, $y_m(t + \mu_m\,T) = y_m(t)$; while if the initial datum $\dot{y}(0)$ implies the complementary *inequality* $\left|\hat{\eta}_m\right| < 1$, this solution is *not periodic*, featuring instead the property $y_m(t + \mu_m\,T) = y_m(t) - 2\mathbf{i}\pi/c_m$ (hence in this case this solution $y_m(t)$ diverges *almost linearly* in the *remote* past and future).

The corresponding results for the system (4.149)—with one or more *vanishing* parameters r_m, identified by their indices m so that

$$r_m = 0 \ \ \text{if} \ \ m = \hat{m}, \quad r_m \neq 0 \ \ \text{if} \ \ m \neq \hat{m} \tag{4.151a}$$

—imply that the solutions $x_n(t)$ are *periodic* with at most (see Remark 2.4.1.2) period $\nu\hat{M}\,T$, where $\hat{M}$ is now the Minimum Common Multiple of the integers k_m with $q_m \neq 0$ and of the N integers μ_m (see (4.149e)) provided the initial data imply,

$$\left|\hat{\eta}_m\right| > 1 \ \ \text{if} \ \ m = \hat{m}\,; \tag{4.151b}$$

while, if for anyone of the indices of type $\hat{m}$ there holds the complementary inequality $\left|\hat{\eta}_{\hat{m}}\right| < 1$, then—see, if need be, Appendix G of the book [20]—in the *remote* past and future some of the coordinates $x_n(t)$ in the *complex x-plane diverge*, indicating the existence also of *scattering*, rather than *periodic*, solutions of the model (4.149) when one or more of the parameters r_m vanishes.

> **Remark 4.3.4.2.** The *N*-body models treated in this Subsection 4.3.4 are *Hamiltonian*: see Remarks 2.4.2.1 and 4.2. ∎

Let us complete this Subsection 4.3.4 by displaying the equations of motion (4.149a) in the case $N = 2$:

$$
\begin{aligned}
\ddot{x}_n = \frac{(-1)^{n+1}}{x_1 - x_2} \Big\{ & 2\,\dot{x}_1\,\dot{x}_2 \\
& + \Big[\mathbf{i}\,(2\,r_1 + 1)\,\rho_1\,\omega\,(\dot{x}_1 + \dot{x}_2) + r_1\,(r_1 + 1)\,(\rho_1\,\omega)^2\,(x_1 + x_2) \\
& + c_1\,[\dot{x}_1\dot{x}_2 - \mathbf{i}\,r_1\,\rho_1\,\omega\,(x_1 + x_2)]^{(2+r_1)/(1+r_1)}\Big]\,x_n \\
& - \mathbf{i}\,(2\,r_2 + 1)\,\rho_2\,\omega\,(\dot{x}_1\,x_2 + \dot{x}_2\,x_1) - r_2\,(r_2 + 1)\,(\rho_2\,\omega)^2\,x_1\,x_2 \\
& - c_2\,(\dot{x}_1\,x_2 + \dot{x}_2\,x_1 - \mathbf{i}\,r_2\,\rho_2\,\omega\,x_1\,x_2)^{(2+r_2)/(1+r_2)}\Big\}, \quad n = 1, 2\,; \tag{4.152a}
\end{aligned}
$$

as well as the special cases of these equations of motion corresponding to the following three assignments of the parameters r_1 and r_2: for $r_1 = -1/2$ and $r_2 = -2$

$$\ddot{x}_n = \frac{(-1)^{n+1}}{x_1 - x_2} \left\{ 2\,\dot{x}_1\,\dot{x}_2 \right.$$

$$\left. + c_1 \left[\dot{x}_1 + \dot{x}_2 + \frac{\mathbf{i}\,\rho_1\,\omega\,(x_1 + x_2)}{2} \right]^3 x_n - c_2 \right\}, \quad n = 1, 2\,; \qquad (4.152\text{b})$$

for $r_1 = -2$ and $r_2 = -1/2$

$$\ddot{x}_n = \frac{(-1)^{n+1}}{x_1 - x_2} \left\{ 2\,\dot{x}_1\,\dot{x}_2 - c_1\,x_n \right.$$

$$\left. - c_2 \left[\dot{x}_1\,x_2 + \dot{x}_2\,x_1 - \frac{\mathbf{i}\,\rho_2\,\omega\,x_1\,x_2}{2} \right]^3 \right\}, \quad n = 1, 2\,; \qquad (4.152\text{c})$$

for $r_1 = -2$ and $r_2 = 0$,

$$\ddot{x}_n = \frac{(-1)^{n+1} \left[2\,\dot{x}_1\,\dot{x}_2 - c_1\,x_n - c_2\,(\dot{x}_1\,x_2 + \dot{x}_2\,x_1)^2 \right]}{x_1 - x_2}, \quad n = 1, 2\,. \qquad (4.152\text{d})$$

4.3.5 An Integrable Hamiltonian N-Body Problem of Goldfish Type, and Its Isochronous Variant

The models treated in Subsection 4.3.5 can be seen as a follow-up—via the approach described in the first part of this Chapter 4—to the findings reported in Subsection 2.4.4. Accordingly, we limit our presentation in Subsection 4.3.5 to a report of the main findings; the interested reader will have no difficulty to provide the corresponding proofs by using the results reported in Subsection 2.4.4, and to analyze the detailed behavior of the solutions of the N-body models of goldfish type exhibited below with their solutions.

Proposition 4.3.5.1. Let the following N-body problem of goldfish type be characterized by the *Newtonian* equations of motion,

$$\ddot{x}_n = \sum_{\ell=1,\,\ell\neq n}^{N} \frac{2\,\dot{x}_n\,\dot{x}_\ell}{x_n - x_\ell} - \left[\prod_{\ell=1,\,\ell\neq n}^{N} (x_n - x_\ell) \right]^{-1} \cdot$$

$$\cdot \sum_{m=1}^{N} \left\{ \left[\rho_m\,(\dot{y}_m)^2\,(y_m)^{-1} + c_m\,(y_m)^{\rho_m} \right] (x_n)^{N-m} \right\}, \qquad (4.153\text{a})$$

where, in the right-hand side, the quantities $\dot{y}_m \equiv \dot{y}_m(t)$ respectively $y_m \equiv y_m(t)$ must be replaced by their expression, (2.5) respectively (2.4), in terms of $x_n \equiv x_n(t)$ and $\dot{x}_n \equiv \dot{x}_n(t)$, while the $2N$ parameters ρ_m and c_m can be *arbitrarily* assigned ($\rho_m \neq 0$, $\rho_m \neq 1$; $c_m \neq 0$). Then the solutions of this model are provided by the N *zeros* of the polynomial (2.3) with its N *coefficients* $y_m \equiv y_m(t)$ given by the following formulas:

$$y_m(t) = y_m(0) \left[1 + (1 - \rho_m) \left\{ \frac{\dot{y}_m(0)\, t}{y_m(0)} + \frac{c_m}{2} \left[y_m(0) \right]^{\rho_m - 1} t^2 \right\} \right]^{1/(1-\rho_m)}. \quad \blacksquare$$

(4.153b)

Remark 4.3.5.1. The N-body model of goldfish type described above, see (4.153), is *Hamiltonian* (see Remark 2.4.4.1), and it is clearly *integrable* (see Remark 4.2). $\blacksquare$

Proposition 4.3.5.2. Let the following N-body problem of goldfish type be characterized by the *Newtonian* equations of motion,

$$\ddot{x}_n = \sum_{\ell=1,\,\ell\neq n}^{N} \frac{2\,\dot{x}_n\,\dot{x}_\ell}{x_n - x_\ell} + \left[\prod_{\ell=1,\,\ell\neq n}^{N} (x_n - x_\ell) \right]^{1} \sum_{m=1}^{N} \left\{ (x_n)^{N-m} \cdot \right.$$

$$\cdot \left[3\,\mathbf{i}\,\rho_m\,\omega\,\dot{y}_m + 2\,(\rho_m\,\omega)^2\,y_m - \left(\frac{r_m + 2}{r_m} \right) (\dot{y}_m)^2\,(y_m)^{-1} \right.$$

$$\left. \left. - c_m\,(y_m)^{(r_m + 2)/r_m} \right] \right\},$$

(4.154a)

where, in the right-hand side, the quantities $\dot{y}_m = \dot{y}_m(t)$ respectively $y_m \equiv y_m(t)$ must be replaced by their expression, (2.5) respectively (2.4), in terms of $x_n \equiv x_n(t)$ and $\dot{x}_n \equiv \dot{x}_n(t)$, while the $2N$ parameters r_m and ρ_m are *arbitrary nonvanishing real rational* numbers ($r_m \neq 0$, $r_m \neq -2$; $\rho_m \neq 0$), the N parameters c_m are essentially *arbitrary* ($c_m \neq 0$), and as usual ω is an *arbitrary nonvanishing real* parameter to which the basic period $T = 2\pi / |\omega|$ is associated. Then the solutions of this model are provided by the N *zeros* of the polynomial (2.3) with its N *coefficients* $y_m \equiv y_m(t)$ given by the following formulas:

$$y_m(t) = y_m(0)\, \exp\left(\mathbf{i}\,r_m\,\rho_m\,\omega\,t\right) \cdot$$

$$\cdot \left[1 - \left(\frac{2}{r_m} \right) \left\{ \frac{\dot{y}_m(0)\,\tau(t)}{y_m(0)} + \frac{c_m}{2} \left[y_m(0) \right]^{2/r_m} [\tau_m(t)]^2 \right\} \right]^{-r_m/2}, \quad (4.154b)$$

where

$$\tau_m(t) = \frac{\exp\left(\mathbf{i}\,\rho_m\,\omega\,t\right) - 1}{\mathbf{i}\,\rho_m\,\omega}. \quad \blacksquare$$

(4.154c)

Remark 4.3.5.2. The model of goldfish type described above, see (4.154), is clearly related to the previous model, see (4.153)—to which it reduces for $\omega = 0$—via the "isochronization trick" detailed in Section 2.4, amounting essentially, for the N coefficients $y_m(t)$, to the change of independent variable from t to $\tau_m(t)$, see (4.154c), and a corresponding appropriate time-dependent rescaling of these variables $y_m(t)$. It is therefore also *Hamiltonian*—with a time-dependent *Hamiltonian*—and it clearly features lots of *periodic* solutions, enough to qualify it as *isochronous*. $\blacksquare$

Remark 4.3.5.3. This *solvable isochronous* and *Hamiltonian* N-body model, characterized by the *Newtonian* equations of motion of goldfish type (4.154a), features a lot of *arbitrary* parameters: the N numbers c_m and the $2N$ *real rational* numbers r_m and ρ_m. ∎

We complete this terse Subsection 4.3.5 with a display of the Newtonian equations of motion (4.154a) in the simplest case with $N = 2$:

$$
\begin{aligned}
\ddot{x}_n = -3\,\mathbf{i}\,\omega\,\dot{x}_n + \frac{(-1)^{n+1}}{x_1 - x_2} &\left\{ -\frac{2}{r_2}\,\dot{x}_1\,\dot{x}_2 \right. \\
&- r_1\,(\rho_1\,\omega)^2\,(x_1 + x_2)\,x_n + r_2\,(\rho_2\,\omega)^2\,x_1\,x_2 \\
&+ \left[\left(\frac{r_1+2}{r_1}\right)\,(\dot{x}_1 + \dot{x}_2)^2\,(x_1+x_2)^{-1} - c_1\,(-x_1 - x_2)^{(r_1+2)/r_1} \right] x_n \\
&\left. - \left(\frac{r_2+2}{r_2}\right) \left[\frac{(\dot{x}_1)^2\,x_2}{x_1} + \frac{(\dot{x}_2)^2\,x_1}{x_2} \right] - c_2\,(x_1\,x_2)^{(r_2+2)/r_2} \right\}.
\end{aligned}
\tag{4.155}
$$

4.4 Another *N*-Body Problem Which Is Isochronous When Its *2N* Parameters Are Arbitrary Rational Numbers

In Section 4.4 we introduce and tersely investigate an N-body problem *solvable* by *algebraic* operations, such as the computation of the N zeros $x_n(t)$ of a known time-dependent polynomial of degree N. It is characterized (see below) by *Newtonian* ("accelerations equal forces") equations describing the motion in the *complex* x-plane of N unit-mass particles interacting *nonlinearly*. These equations of motion are *rotation-invariant* but *not* translation-invariant. They feature $2N$ *arbitrary* (coupling) constants; as shown below, when these parameters are restricted to be *real* and *rational* the model is *isochronous*, its solutions yielded by *generic* initial data being then *all periodic* with a *fixed* period (while many solutions are also periodic with a *smaller* period).

This model obtains from the key formula (4.79) with

$$
\ddot{y}_m = \frac{c_m\,(\dot{y}_m)^2}{y_m} + \mathbf{i}\,b_m\,\omega\,\dot{y}_m,
\tag{4.156}
$$

where the $2N$ parameters c_m, b_m are *arbitrary* (but see below). Hence its equations of motion read as follows:

$$
\begin{aligned}
\ddot{x}_n = \sum_{\ell=1,\,\ell \neq n}^{N} \left(\frac{2\,\dot{x}_n\,\dot{x}_\ell}{x_n - x_\ell} \right) - \left[\prod_{\ell=1,\,\ell \neq n}^{N} (x_n - x_\ell) \right]^{-1} \cdot \\
\cdot \sum_{m=1}^{N} \left\{ \left[\frac{c_m\,(\dot{y}_m)^2}{y_m} + \mathbf{i}\,b_m\,\omega\,\dot{y}_m \right] (x_n)^{N-m} \right\},
\end{aligned}
\tag{4.157}
$$

where the $2N$ quantities $y_m(t)$ respectively $\dot{y}_m(t)$ must be expressed in terms of the N particle coordinates $x_n(t)$ and their time-derivatives $\dot{x}_n(t)$ by the *explicit* formulas (2.4) respectively (2.5).

Remark 4.4.1. The equations of motion (4.156) can be *formally* obtained from the Hamiltonian

$$H(\vec{p}, \vec{y}) = \sum_{m=1}^{N} h_m(p_m, y_m),$$ (4.158a)

$$h_m(p, y) = \left[a_m y^{c_m} + \left(\frac{\mathbf{i} b_m \omega}{1 + c_m} \right) y \right] p,$$ (4.158b)

where the N quantities y_m (components of the N-vector $\vec{y}$) play the role of *canonical variables*, the N quantities p_m (components of the N-vector $\vec{p}$) are the corresponding *canonical momenta*, and the N *nonvanishing* constants a_m can be assigned *arbitrarily*. But this Hamiltonian—*linear* in the canonical momenta—is *not* acceptable, since the first set of equations implied by it reads as follows,

$$\dot{y}_m = \frac{d h_m(p_m, y_m)}{d p_m} = a_m (y_m)^{c_m} + \left(\frac{\mathbf{i} b_m \omega}{1 + c_m} \right) y_m,$$ (4.158c)

and although these ODEs do imply the second-order ODEs (4.156), they are consistent with only a *subclass* of solutions of these second-order ODEs, namely only with those characterized by initial data $y_m(0)$ and $\dot{y}_m(0)$ satisfying the following *restriction* implied by (4.158c):

$$\dot{y}_m(0) = \alpha_m \left[y_m(0) \right]^{c_m} + \left(\frac{\mathbf{i} b_m \omega}{1 + c_m} \right) y_m(0). \ \blacksquare$$ (4.158d)

The solution of the initial-value problem of the equations of motion (4.156) reads as follows: if $c_m = 1$, then

$$y_m(t) = y_m(0) \exp \left\{ \frac{\dot{y}_m(0)}{y_m(0)} \left[\frac{\eta_m(t) - 1}{\mathbf{i} b_m \omega} \right] \right\} ;$$ (4.159a)

if $c_m \neq 1$, then

$$y_m(t) = y_m(0) \left[\varphi_m(t) \right]^{1/(1 - c_m)},$$

$$\varphi_m(t) = 1 + \frac{(1 - c_m) \dot{y}_m(0)}{y_m(0)} \left[\frac{\eta_m(t) - 1}{\mathbf{i} b_m \omega} \right]$$

$$= \frac{(1 - c_m) \dot{y}_m(0)}{\mathbf{i} b_m \omega y_m(0)} [\eta_m(t) - \bar{\eta}_m],$$ (4.159b)

with (above and below)

$$\eta_m(t) = \exp(\mathbf{i} b_m \omega t), \quad \bar{\eta}_m = 1 - \frac{\mathbf{i} b_m \omega y_m(0)}{(1 - c_m) \dot{y}_m(0)}.$$ (4.159c)

Hence the solution $x_n(t)$ of the N-body model (4.157) is provided by the N zeros of the polynomial (2.3) the N *coefficients* $y_m(t)$ of which are given by these formulas (4.159) with the $2N$ data $y_m(0)$ respectively $\dot{y}_m(0)$ defined in terms of the $2N$ initial data $x_n(0)$ and $\dot{x}_n(0)$ by the formulas (2.4) respectively (2.5) at $t = 0$.

These formulas imply that if ω is *real* and *nonvanishing* and the $2N$ parameters c_m and b_m are *all real rational* numbers, then—generically, i.e., for *all complex* initial data $y_m(0)$ and $\dot{y}_m(0)$ except for a *subset* of such data having *vanishing* measure—the time evolution of all the coefficients $y_m(t)$ is *periodic* with a *common period* which is an *integer multiple* of the basic period $T = 2\pi / |\omega|$; hence *all* solutions of the N-body problem (4.157) are also *completely periodic* with a *common period* which is as well some *integer multiple* of the basic period $T = 2\pi / |\omega|$ (see Section 2.3). The *isochronicity* property of this N-body model—as advertised in the title of this Section 4.4—is thereby confirmed.

> **Remark 4.4.2.**   The model introduced and investigated in Section 4.4 is a *generalization* of models treated in previous sections, see in particular Section 4.1 and Subsection 4.3.2. ∎

> **Remark 4.4.3.**   The equations of motion (4.157) are *invariant* under the rescaling $x_n(t) \Rightarrow c\, x_n(t)$, with c an *arbitrary (nonvanishing) complex constant*. Hence in particular this N-body problem (4.157) is invariant under the rescaling $x_n(t) \Rightarrow \exp(\mathbf{i}\,\theta)\, x_n(t)$, with θ an *arbitrary time-independent* angle, amounting to a common *rotation* of the system around the origin of the *complex* x-plane. This suggests the possible convenience to change the dependent variables from the *complex* numbers $x_n(t) = \text{Re}\,[x_n(t)] + \mathbf{i}\,\text{Im}\,[x_n(t)]$ to the *real* two-vectors $\vec{r}_n(t)$ with components $u_n(t) = \text{Re}\,[x_n(t)]$, $v_n(t) = \text{Im}\,[x_n(t)]$, implying that the particles then move in the *real* Cartesian uv-plane rather than in the *complex* x-plane, according to *rotation-invariant* Newtonian equations of motion. The interested reader can reformulate the equations of motion appropriate to this new *real two-dimensional* context by taking advantage of the "dictionary" provided below, see Appendix. ∎

In addition to the case with *rational* parameters c_m and b_m discussed above, some other, more general, assignments of these parameters deserve to be discussed inasmuch as they allow to identify a *remarkable* behavior of the corresponding N-body model (4.157). For instance, let us consider the case when the parameters c_m are again *all real rational* numbers, but only *some* of the parameters b_m are as well *real* and *rational* while *some others* are *arbitrary complex* numbers featuring a *positive imaginary part*: so that for some values of the index m in the range $1 \leq m \leq N$ there holds the property $\text{Im}\,[b_m] > 0$ (with $\text{Re}\,[b_m]$ unrestricted), while for *all other* values of the index m the parameters b_m are *real nonvanishing rational* numbers. It is then clear—see the more detailed discussion below—that the corresponding N-body problems in (4.157) are *asymptotically isochronous*, namely their *generic* solutions are, in the *remote* future, *all completely periodic* (with a period which is an *integer multiple* of the basic period $T = 2\pi/|\omega|$), up to contributions which vanish exponentially as $t \to +\infty$. It is thus seen that the N-body model (4.157) also includes a vast class of *asymptotically isochronous* cases (for this notion and additional examples of this phenomenology see for instance [54] and Chapter 6 of [27]).

As indicated above, the qualitative properties of the N-body problem (4.157) discussed above characterize their *generic* solutions: excluded as *nongeneric* are the solutions running into *singularities*. Let us turn now to a terse discussion of the possible emergence of such *singular* solutions.

The equations of motion (4.157) are quite *nonlinear*. Generally the solutions of such nonlinear dynamical systems display a quite complicated time evolution (including "deterministic chaos") and a large collection of nontrivial *singularities*. The *solvable* character of this N-body system allows instead a complete understanding of its *overall behavior over time* and of the *singularity structure* of its solutions.

For *nonlinear* (systems of) ODEs there are generally two possible types of singularities: those that are immediately visible from the structure of the equations of motion, and those which are instead due to the *nonlinear* character of the equations of motion but are not immediately evident from the structure of these equations. Both types are generally present in this model (4.157).

The first type of singularities have a quite evident "physical" connotation: they are associated with *particle collisions*. As discussed above and as is evident from the solution of this model, these phenomena do *not* cause any blow-up of the solutions but rather a "loss of identity" of the colliding particles and a *singularity* of the second time-derivative of their coordinates $x_n(t)$ (see the right-hand side of the equations of motion (4.157)). It is also evident that—at least in the *complex* context with *periodic* motions (see above)—solutions featuring this phenomenology are *not generic*: indeed, if a solution yields particle collisions, any *generic* modification of the initial data will cause the solution to avoid the collision—because in the *complex* context the N zeros of a *generic* polynomial of degree N are *distinct*, i.e., *all different from each other*.

The second type of singularities can be read directly from the explicit expressions (4.159) of the coefficients $y_m(t)$ of the polynomial (2.3)—the *zeros* $x_n(t)$ of which yield the solutions of the equations of motion (4.157), implying that such singularities can only emerge from singularities of these coefficients $y_m(t)$. The equations of motion (4.157), and even more evidently the explicit expressions (4.159) of the quantities $y_m(t)$, show that the emergence of *singularities* can only be due to the possibility that some coefficient $y_m(t)$ *vanish* over its time evolution; note that the very structure of the N-body model (4.157) implies that the *initial* data of the model must *exclude* the possibility that any one of the quantities $y_m(t)$ vanish *initially*, $y_m(0) \neq 0$ (unless the corresponding parameter c_m vanishes, $c_m = 0$: see (4.156)). It is evident that no such singularities exist if *all* the parameters c_m have the following values: $c_m = 1$ (see (4.159a)), or $c_m = 1 - 1/k_m$ hence $1/(1 - c_m) = k_m$ (see (4.159b)) with k_m an *arbitrary positive integer*, $k_m \geq 1$. Otherwise the coefficient $y_m(t)$—considered as function of the *complex* variable $\eta_m(t) = \exp(\mathbf{i}\, b_m\, \omega\, t)$, see (4.159c)—will feature a *singularity*—a branch point or a pole, as the case may be, see (4.159b)—at the value

$$\eta_m(t) = \bar{\eta}_m = 1 - \frac{\mathbf{i}\, b_m\, \omega\, y_m(0)}{(1 - c_m)\, \dot{y}_m(0)}\, ; \qquad (4.160a)$$

but this singularity will be encountered by $y_m(t)$ over its time evolution—assuming that the parameter b_m is *real*, so that $|\eta_m(t)| = 1$, see (4.159c)—only if the initial data $y_m(0)$ and $\dot{y}_m(0)$ satisfy the *equality*

$$|\bar{\eta}_m| \equiv \left| 1 - \frac{\mathbf{i}\, \omega\, b_m\, y_m(0)}{(1 - c_m)\, \dot{y}_m(0)} \right| = 1. \qquad (4.160b)$$

Hence the solutions $x_n(t)$ of the N-body problem (4.157) will be *singular* only if at least for one value of m—with $c_m \neq 1$ and $c_m \neq 1 - 1/k_m$ (with k_m an *arbitrary positive integer*)—there holds this *equality*, (4.160b): implying that the corresponding initial data are, again, *not generic*.

We therefore conclude that—at least if the parameters b_m are *all real* and *rational* (so that the solutions are *periodic*, see above)—the *generic* solutions of the N-body problem (4.157) are *nonsingular*.

Let us now discuss in somewhat more detail the behavior of the *generic (nonsingular)* solutions of this N-body problem (4.157), always restricting for simplicity our treatment to the case when the $2N$ parameters c_m and b_m are *all real rational* numbers:

$$c_m = \frac{\hat{c}_m}{\check{c}_m}, \quad b_m = \frac{\hat{b}_m}{\check{b}_m}, \qquad (4.161)$$

where $\hat{c}_m$ and $\check{c}_m$—and likewise $\hat{b}_m$ and $\check{b}_m$—are *nonvanishing coprime integers* (for definiteness let us assume the denominators $\check{c}_m$ and $\check{b}_m$ to be *positive*, $\check{c}_m \geq 1$, $\check{b}_m \geq 1$). It is then clear that the quantity $\eta(t) = \exp(\mathbf{i}\, b_m\, \omega\, t)$ is *periodic* with period $T_m^{(b)} = 2\pi/|b_m\, \omega|$, hence the quantity $\varphi_m(t)$ (see (4.159b)) is *periodic* with the same period $T_m^{(b)}$ and $y_m(t)$ is clearly as well periodic with the same period $T_m^{(b)}$ if $c_m = 1$ hence $\hat{c}_m = \check{c}_m$ (see (4.159a) and (4.161)) or $c_m = 1 - 1/k_m$ (implying $1/(1 - c_m) = k_m = \check{c}_m/(\hat{c}_m - \check{c}_m)$; see (4.159b) and (4.161)) with k_m an *arbitrary integer*. For other (*rational*) values of c_m the coefficient $y_m(t)$ is also *periodic* in t (see (4.159b)), but the value of the period is then different depending whether the branch point (of order $\hat{c}_m - \check{c}_m$)—occurring at the value $\bar{\eta}_m$ (see (4.160a)) of the complex variable η at which the quantity φ_m (see (4.159b)) vanishes—is *outside* or *inside* the circle $\tilde{C}$—centered at the origin and of unit radius in the *complex* plane— traveled by the quantity $\eta_m(t)$, see (4.159c), over its period $T_m^{(b)}$. Hence we conclude that if the initial data $y_m(0)$ and $\dot{y}_m(0)$ imply that $|\bar{\eta}_m| > 1$, then the periodicity of $y_m(t)$ is unaffected, i.e. $y_m(t)$ is periodic with period $T_m^{(b)}$,

$$y_m\left(t + T_m^{(b)}\right) = y_m(t), \quad T_m^{(b)} = 2\pi \left|\check{b}_m/\omega\right|, \tag{4.162a}$$

while if instead $|\bar{\eta}_m| < 1$, then $y_m(t)$ is *periodic* with the period,

$$y_m\left(t + T_m^{(bc)}\right) = y_m(t),$$

$$T_m^{(bc)} = 2\pi \left|\check{b}_m \left(\hat{c}_m - \check{c}_m\right)/\omega\right| = \left|\hat{c}_m - \check{c}_m\right| T_m^{(b)}. \tag{4.162b}$$

Note that this implies that the period of $y_m(t)$ depends on the initial data $y_m(0)$ and $\dot{y}_m(0)$; however, for *arbitrary* initial data, $y_m(t)$ is certainly *periodic* with period $T_m^{(bc)}$ since $\left|\hat{c}_m - \check{c}_m\right|$ is a *positive integer* hence $T_m^{(bc)}$ is an *integer multiple* of $T_m^{(b)}$.

Next, let us discuss what this entails for the time evolution of the solutions $x_n(t)$, as implied by the fact that they are the N *zeros* of the polynomial (2.3) the N *coefficients* $y_m(t)$ of which are *periodic* as detailed above. It is then evident that the periodicity of *all generic* (hence *nonsingular*) solutions of the N-body problem (4.157) are—in this case with *rational* parameters c_m and b_m, see (4.161)—*periodic*, $x_n(t + T) = x_n(t)$, $n = 1, \ldots, N$, with at most (see Remark 2.4.1.2) period

$$T = \nu_{\text{Max}}(N) \ \underset{m=1,\ldots,N}{\text{MinimumCommonMultiple}}\left[\left|\check{b}_m \left(\hat{c}_m - \check{c}_m\right)\right|\right] \frac{2\pi}{|\omega|}, \tag{4.163}$$

where $\nu_{\text{Max}}(N)$ is a *positive integer* the significance of which is explained in Section 2.3. This demonstrates the *isochronous* character of this N-body model. As implied by the above discussion there are *generic* sets of initial data $x_n(0)$, $\dot{x}_n(0)$ which yield solutions *periodic* with periods that are *integer submultiples* of this period T, see (4.163) (and clearly the sets of initial data yielding solutions with *different* periods are *separated* by initial data yielding *collisions*: see Remark 2.3.1).

Another interesting case of the N-body problem (4.157) that deserves to be tersely discussed is that characterized by N parameters c_m, which are (again) *real* and *rational* while only a (*nonempty*) subset of the N parameters b_m are *real* and *rational* numbers while the elements b_m of the (also *nonempty*) complementary subset are *complex* numbers the *real* parts of which are *arbitrary* while their (*nonvanishing*) *imaginary* parts are *arbitrary positive* numbers β_m:

$$c_m = \frac{\hat{c}_m}{\check{c}_m}, \quad b_m = \text{Re}\left[b_m\right] + \mathbf{i}\,\beta_m, \quad \beta_m \geq 0, \tag{4.164a}$$

with, as above, $\hat{c}_m$, $\check{c}_m$ *real nonvanishing coprime integers* and with some of the N *real* numbers β_m, say $\beta_m = \beta_{\check{m}}$, *vanishing*, $\beta_{\check{m}} = 0$, and the corresponding *(real) numbers* $b_{\check{m}}$ *rational*, $b_{\check{m}} = \hat{b}_{\check{m}}/\check{b}_{\check{m}}$ (with $\hat{b}_{\check{m}}$ and $\check{b}_m$ *nonvanishing coprime integers*), while some others of the N *real* numbers β_m, say $\beta_m = \beta_{\hat{m}}$, are *positive real numbers*, $\beta_{\hat{m}} > 0$. The corresponding polynomial (2.3) is then *asymptotically isochronous*, i.e.

$$p_N\left(z; \vec{y}(t), \tilde{x}(t)\right) - \tilde{p}_N\left(z; t\right) \to 0 \quad \text{as} \quad t \to \infty, \tag{4.164b}$$

where $\tilde{p}_N\left(z; t\right)$ is *periodic* in t (while $p_N\left(z; \vec{y}(t); \tilde{x}(t)\right)$ is *not* periodic): in self-evident notation,

$$\tilde{p}_N\left(z; t\right) = z^N + \sum_{\hat{m}=1}^{N}\left[y_{\hat{m}}\left(\infty\right) z^{N-\hat{m}}\right] + \sum_{\check{m}=1}^{N}\left[y_{\check{m}}(t) z^{N-\check{m}}\right], \tag{4.164c}$$

with $y_{\check{m}}(t)$ defined by (4.159) (with m replaced by $\check{m}$) and $y_{\hat{m}}\left(\infty\right)$ also defined by (4.159) (with m replaced by $\hat{m}$ and with $\eta_{\hat{m}}(t)$ replaced by $\eta_{\hat{m}}\left(\infty\right) = 0$).

The solutions of the N-body problem (4.157) with these parameters c_m and b_m, see (4.164a), are then as well *asymptotically periodic*,

$$x_n\left(t + \nu_{\text{Max}}\left(N\right) \tilde{T}\right) - x_n(t) \to 0 \quad \text{as} \quad t \to \infty, \tag{4.164d}$$

where $\tilde{T}$ is the period of the periodic polynomial $\tilde{p}_N\left(z; t\right)$, see (4.164c) and the treatment detailed above—and for the significance of the *positive integer* $\nu_{\text{Max}}(N)$, see again Section 2.3.

We leave to the interested reader an analysis of the case—more general than that treated above—with the parameters c_m again all *real* and the parameters b_m being again *complex* but now some of them having a *negative imaginary* part, implying that the corresponding quantities $\eta_m(t) = \exp\left(\mathbf{i}\, b_m\, \omega\, t\right)$ diverge as $t \to \infty$. In that case—provided $c_m < 1$ for at least some of the corresponding values of m: see (4.159b)—the solutions of the N-body model (4.157) will feature some particles which escape from the *finite* part of the *complex* x-plane as $t \to \infty$: see, for a relevant analysis, see Appendix G entitled "Asymptotic behavior of the zeros of a polynomial whose coefficients diverge exponentially" in the book [20].

Let us end this Section 4.4 by displaying the equations of motion of the N-body model considered herein, see (4.157), in the simplest case with $N = 2$:

$$\ddot{x}_n = \frac{(-1)^{n+1}}{x_1 - x_2}\left\{2\,\dot{x}_1\,\dot{x}_2 + \left[\frac{c_1\,(\dot{x}_1 + \dot{x}_2)^2\,x_n}{x_1 + x_2} - \frac{c_2\,(\dot{x}_1\,x_2 + x_1\,\dot{x}_2)^2}{x_1\,x_2}\right]\right.$$
$$\left. + \mathbf{i}\,\omega\,\left[b_1\,(\dot{x}_1 + \dot{x}_2)\,x_n - b_2\,(\dot{x}_1\,x_2 + x_1\,\dot{x}_2)\right]\right\}, \quad n = 1, 2. \tag{4.165}$$

Again—see above Remark 4.4.3—the reformulation of this model in terms of motions of *real* 2-vectors in the Cartesian plane rather than points in the *complex* plane is left as an instructive exercise for the interested reader, who will find useful for this purpose the "dictionary"—facilitating this transition—provided below, see Appendix.

4.5 Two Classes of Solvable *N*-Body Problems of Goldfish Type

In Section 4.5, two classes of *N*-body problems of goldfish type are investigated, with *N* an *arbitrary positive integer* ($N \geq 2$). These models are characterized by *nonlinear* Newtonian ("accelerations equal forces") equations of motion describing *N* unit-mass point-particles moving in the *complex x-plane*. These highly *nonlinear* equations feature *many arbitrary* coupling constants, yet they can be *solved* by *algebraic* operations. Some of these *N*-body problems are *isochronous*, their *generic* solutions being *all completely periodic* with an overall period *T* independent of the initial data (but quite a few of these solutions are actually periodic with *smaller* periods T/p with *p* a *positive integer*); other models are *isochronous* for an *open* region of initial data, while the motions for other initial data are *not periodic*, featuring instead *scattering* phenomena with some of the particles incoming from, or escaping to, *infinity* in the *remote* past and future. The findings reported in Section 4.5 are based on the results described in Subsection 2.4.5.

The *first* class of *solvable N*-body problems is characterized by the Newtonian equations of motion

$$
\ddot{x}_n = \sum_{\ell=1,\,\ell\neq n}^{N} \left[\frac{2\,\dot{x}_n\,\dot{x}_\ell}{x_n - x_\ell} \right] - \left[\prod_{\ell=1,\,\ell\neq n}^{N} (x_n - x_\ell)^{-1} \right] \cdot
$$

$$
\cdot \sum_{m=1}^{N} \left\{ \left[\mathbf{i}\,(2\,r_m + 1)\,\rho_m\,\omega\,\dot{y}_m + r_m\,(r_m + 1)\,(\rho_m\,\omega)^2\,y_m \right. \right.
$$

$$
\left. \left. + c_m\,(\dot{y}_m - \mathbf{i}\,r_m\,\rho_m\,\omega\,y_m)^{(1+2r_m)/(1+r_m)}\,(y_m)^{(1-r_m)/r_m} \right] (x_n)^{N-m} \right\}. \quad (4.166a)
$$

Here ω is an *arbitrary nonvanishing real* number to which we associate as usual the basic period

$$
T = \frac{2\,\pi}{|\omega|}, \quad (4.166b)
$$

the *N* coupling constants c_m are *arbitrary complex* numbers, and the quantities $y_m(t)$ respectively $\dot{y}_m(t)$ are expressed in terms of the coordinates $x_n(t)$ and $\dot{x}_n(t)$ via the standard formulas (2.4) and (2.5). As for the 2*N* parameters r_m and ρ_m, we hereafter assume them to be *arbitrary rational* numbers,

$$
r_m = \frac{q_m}{k_m}, \quad \rho_m = \frac{\lambda_m}{\mu_m} \quad (4.166c)
$$

where q_m and k_m are two *arbitrary coprime integers* ($k_m \geq 1$; if $c_m \neq 0$ then $q_m \neq 0$ and $q_m \neq -k_m$) and likewise λ_m and μ_m are two *arbitrary coprime integers* ($\mu_m \geq 1$); and we moreover define, for future reference, the *rational* number

$$s_m = r_m \, \rho_m = \frac{q_m \, \lambda_m}{k_m \, \mu_m} = \frac{\hat{s}_m}{\check{s}_m} \tag{4.166d}$$

where $\hat{s}_m$ and $\check{s}_m$ are two *coprime integers* ($\check{s}_m \geq 1$). The quantities $\dot{y}_m \equiv \dot{y}_m(t)$ and $y_m \equiv y_m(t)$ in the right-hand side of (4.166a) are of course supposed to be replaced by their expressions in terms of $x_n \equiv x_n(t)$ and $\dot{x}_n \equiv \dot{x}_n(t)$, see (2.4) and (2.5).

The *solvable* character of this N-body problem is demonstrated by the following:

Proposition 4.5.1. The N *complex* coordinates $x_n(t)$ providing the solution at time t of the initial-value problem of the Newtonian equations of motion (4.166a) are the N *zeros* of the monic polynomial (see (2.3)) the N coefficients $y_m(t)$ of which are given, in terms of the initial data $x_n(0)$, $\dot{x}_n(0)$, by the following formulas:

$$y_m(t) = \tilde{y}_m(\eta_m), \quad \eta_m \equiv \eta_m(t) = \exp(\mathbf{i}\,\omega_m \, t), \quad \omega_m = \rho_m \, \omega \tag{4.167a}$$

$$\tilde{y}_m(\eta) = \left(\frac{a_m \, \eta}{C_m}\right)^{r_m} \left[\left(\frac{\bar{\eta}_{m1} - \bar{\eta}_{m2}}{\bar{\eta}_{m1} - \eta}\right)^{1/r_m} - 1\right]^{-r_m}, \tag{4.167b}$$

$$a_m = \left[\dot{y}_m(0) - \mathbf{i}\,r_m\,\omega_m\,y_m(0)\right]^{1/(1+r_m)} - C_m \left[y_m(0)\right]^{1/r_m}, \tag{4.167c}$$

$$b_m = (a_m)^{-1} \left\{a_m \left[y_m(0)\right]^{-1/r_m} + C_m\right\}^{-r_m}, \tag{4.167d}$$

$$\bar{\eta}_{m1} = 1 - \mathbf{i}\,\omega_m\,b_m, \quad \bar{\eta}_{m2} = \bar{\eta}_{m1} + \mathbf{i}\,\omega_m\,(a_m)^{-1}\,(C_m)^{-r_m}, \tag{4.167e}$$

$$C_m = \frac{c_m \, r_m}{1 + r_m}, \tag{4.167f}$$

where the quantities $y_m(0)$ and $\dot{y}_m(0)$ are given in terms of the initial data $x_n(0)$, $\dot{x}_n(0)$ by the formulas (2.4) and (2.5) at $t = 0$, and the determinations of the *rational* roots are implied by the requirement that these formulas be valid at the initial time $t = 0$ and thereafter by *continuity* in t; while if $c_m = 0$ the expression (4.167) must be replaced by

$$y_m(t) = y_m(0) + \dot{y}_m(0) \left[\frac{\exp(\mathbf{i}\,r_m\,\omega_m\,t) - 1}{\mathbf{i}\,r_m\,\omega_m}\right]. \quad \blacksquare \tag{4.168}$$

The proof of this Proposition 4.5.1 need not be detailed: it is a rather immediate consequence of the key formula (4.79) and of the results reported in Subsection 2.4.5, see in particular equation (2.57) (of course with $y(t)$ replaced by $y_m(t)$ and ω replaced by $\omega_m = \rho_m \, \omega$).

The time evolution of this *first* class of *solvable* N-body problems is tersely described by the following

Remark 4.5.1. If neither r_m nor $1/r_m$ are *integers*, for *generic* initial data the function $y_m(\eta_m)$, see (4.167), features in the *complex* τ-plane, two *rational* branch points at $\eta = \bar{\eta}_{m1}$ respectively at $\eta = \bar{\eta}_{m2}$. Hence the *periodicity* of $y_m(t)$ as function of t, see (4.167), is determined by the location in the *complex* η-plane of these two branch points with respect to the circle $\tilde{C}_m$ centered at the origin, $\eta = 0$, and with unit radius, on which rotates the point $\eta_m(t) = \exp(\mathbf{i}\omega_m t)$ as function of the time t—making a full round in a time $T_{\omega_m} = 2\pi / |\omega_m|$. If *both* branch points are located *outside* the circle $\tilde{C}_m$—and the condition on the initial data determining this outcome is clearly validity of *both inequalities*

$$|\bar{\eta}_{m1}| > 1, \quad |\bar{\eta}_{m2}| > 1 \tag{4.169a}$$

with $\bar{\eta}_{m1}$ respectively $\bar{\eta}_{m2}$ defined in terms of the initial data by (4.167c-f)—then clearly $y_m(t) \equiv \tilde{y}_m(\eta_m(t))$ as function of t is *periodic* with the same period as $\eta_m(t) = \exp(\mathbf{i}\,\omega_m\,t)$,

$$y_m(t + T(\mu_m)) = y_m(t), \quad T(\mu_m) = \mu_m\,T, \quad T = 2\pi/\omega, \tag{4.169b}$$

hence $y_m(t)$, see (4.167), is periodic with, at most (see Remark 2.4.1.2), the period (see (4.166c) and (4.166d))

$$T(S_m) = S_m\,T, \quad S_m = \text{MinimumCommonMultiple}\left[\mu_m, \check{s}_m\right], \tag{4.169c}$$

$$y_m(t + T(S_m)) = y_m(t). \tag{4.169d}$$

As a consequence—if the inequalities (4.169a) are valid for *all* values of m in the range from 1 to N—then the time-dependent monic polynomial with the N *coefficients* $y_m(t)$ (see (2.3)) is *periodic* in t with at most (see Remark 2.4.1.2) period

$$T(M) = M\,T, \quad M = \underset{\bar{m}=1,2,\dots,N}{\text{MinimumCommonMultiple}}\left[\mu_{\bar{m}}, \check{s}_{\bar{m}}\right], \tag{4.169e}$$

where the Minimum Common Multiple must be evaluated for the values of $m = \bar{m}$ such that $c_{\bar{m}} \neq 0$; hence the N coordinates $x_n(t)$—being the N *zeros* of this polynomial—are *periodic* with at most (see Remark 2.4.1.2) a period $\nu\,M\,T$ where ν is a *positive integer* in the range from 1 to ν_{Max}, the significance of which is explained in Section 2.3. ∎

It is also evident that the solutions $x_n(t)$ are periodic in t with a period that is a *positive integer multiple* $\tilde{M}$ of the basic period T even if the initial data imply instead that some of the inequalities (4.169a) are reversed; but in these cases the determination of the value of $\tilde{M}$ requires an analysis of the structure of the Riemann surfaces associated to the functions

$$\varphi_m(\eta) = \left\{\left(\frac{\bar{\eta}_{m1} - \bar{\eta}_{m2}}{\bar{\eta}_{m1} - \eta}\right)^{1/r_m} - 1\right\}^{-r_m} \tag{4.170}$$

of the *complex* variable η, and consequently of the periodicity of $\varphi_m(t) \equiv \varphi_m(\eta_m(t))$ as a function of the *real* variable t ("time") when $\eta_m(t) = \exp(\mathbf{i}\,\omega_m\,t)$ so that $\eta_m(t)$ travels on the circle $\tilde{C}_m$ defined above; and then an analysis of the resulting periodicity of the solutions $x_n(t)$ analogous to that made above. We leave this task to the interested reader.

Let us complete this discussion of the *first* class of *solvable N*-body problems treated in Section 4.5 by displaying the equations of motion (4.166a) in the (simplest) $N = 2$ case:

$$\ddot{x}_n = \frac{(-1)^{n+1}}{x_1 - x_1} \Big\{ 2\,\dot{x}_1\,\dot{x}_2$$
$$+ x_n \Big[\mathbf{i}\,(2\,r_1 + 1)\,\rho_1\,\omega\,(\dot{x}_1 + \dot{x}_2) + r_1\,(r_1 + 1)\,(\rho_1\,\omega)^2\,(x_1 + x_2)$$
$$- c_1\,[-\dot{x}_1 - \dot{x}_2 + \mathbf{i}\,r_1\,\rho_1\,\omega\,(x_1 + x_2)]^{(1+2r_1)/(1+r_1)}\,(-x_1 - x_2)^{(1-r_1)/r_1} \Big]$$
$$- \mathbf{i}\,(2\,r_2 + 1)\,\rho_2\,\omega\,(\dot{x}_1\,x_2 + \dot{x}_2\,x_1) - r_2\,(r_2 + 1)\,(\rho_2\,\omega)^2\,x_1\,x_2$$
$$- c_2\,(\dot{x}_1\,x_2 + x_1\,\dot{x}_2 - \mathbf{i}\,r_2\,\rho_2\,\omega\,x_1\,x_2)^{(1+2r_2)/(1+r_2)}\,(x_1\,x_2)^{(1-r_2)/r_2} \Big\},$$
$$n = 1, 2;$$

(4.171)

and by also displaying the Newtonian equations of motion (4.166a) in the special case with $r_m = -1/2$, $\rho_m = 1$ for all values of m and $c_m = 0$ for $m > 1$, when (via (2.6)) they read

$$\ddot{x}_n = -\frac{1}{4}\,\omega^2\,(x_n)^N + \sum_{\ell=1,\,\ell \neq n}^{N} \left[\frac{2\,\dot{x}_n\,\dot{x}_\ell}{x_n - x_\ell} \right]$$
$$+ c_1 \left[\prod_{\ell=1,\,\ell \neq n}^{N} (x_n - x_\ell)^{-1} \right] \left[\sum_{\ell=1}^{N} (x_\ell) \right]^{-3} (x_n)^{N-1}.$$

(4.172)

The *second* class of *solvable N*-body problems that we report in Section 4.5 is characterized by the Newtonian equations of motion

$$\ddot{x}_n = \sum_{\ell=1,\,\ell \neq n}^{N} \left[\frac{2\,\dot{x}_n\,\dot{x}_\ell}{x_n - x_\ell} \right] - \left[\prod_{\ell=1,\,\ell \neq n}^{N} (x_n - x_\ell)^{-1} \right] \cdot$$
$$\cdot \sum_{m=1}^{N} \left\{ \left[3\,\mathbf{i}\,\omega_m\,\dot{y}_m + 2\,(\omega_m)^2\,y_m + c_m\,(\dot{y}_m - \mathbf{i}\,\omega_m\,y_m)\,y_m \right] (x_n)^{N-m} \right\},$$

(4.173)

where, in the right-hand side, the $2N$ parameters ω_m and c_m are *a priori arbitrary* (but see below), and the $2N$ quantities $\dot{y}_m \equiv \dot{y}_m(t)$ respectively $y_m \equiv y_m(t)$ should of course be replaced by their expressions (2.5) respectively (2.4) in terms of the coordinates $x_n \equiv x_n(t)$ and $\dot{x}_n \equiv \dot{x}_n(t)$. As implied by the results reported in the last part of Subsection 2.4.5 (that dealing with the case $\alpha = \beta = 1$; see (2.64)), its *solvability* is demonstrated by the following

Proposition 4.5.2. The N *complex* coordinates $x_n(t)$ providing the solution at time t of the initial-value problem of the Newtonian equations of motion (4.173) are the N *zeros* of the monic polynomial (see (2.3)) the N coefficients $y_m(t)$ of which are given, in terms of the initial data $x_n(0)$, $\dot{x}_n(0)$, by the following formulas:

$$y_m(t) = \exp\left(\mathbf{i}\,\omega_m\,t\right)\,\gamma_m\left(\tau_m(t)\right), \quad \tau_m(t) = \frac{\exp\left(\mathbf{i}\,\omega_m\,t\right) - 1}{\mathbf{i}\,\omega_m}, \tag{4.174a}$$

$$\gamma_m\left(\tau\right) = \frac{y_m(0) - a_m\,\tanh\left(\frac{a_m\,c_m\,\tau}{2}\right)}{1 - \frac{y_m(0)}{a_m}\,\tanh\left(\frac{a_m\,c_m\,\tau}{2}\right)}, \tag{4.174b}$$

$$(a_m)^2 = \left[y_m(0)\right]^2 - \frac{2\left[\dot{y}_m(0) - \mathbf{i}\,\omega_m\,y_m(0)\right]}{c_m}, \tag{4.174c}$$

where of course the $2N$ quantities $y_m(0)$ respectively $\dot{y}_m(0)$ should be replaced by their expressions (2.5) respectively (2.4) in terms of the coordinates $x_n(t)$ and $\dot{x}_n(t)$ at $t = 0$. ∎

Remark 4.5.2. It is clear that, if *all* the parameters ω_m are *rational multiples* of the same *real nonvanishing* parameter ω,

$$\omega_m = \rho_m\,\omega, \quad \rho_m = \frac{\lambda_m}{\mu_m} \tag{4.175}$$

(see (4.166c)), the N-body problem (4.173) is *isochronous*. A more detailed analysis of its behavior—analogous to those reported above—is left to the interested reader, who may to this end take advantage of the findings reported in the last part of Subsection 2.4.5. ∎

Let us complete this discussion of the *second* class of *solvable N*-body problems treated in Section 4.5 by displaying the equations of motion (4.173) in the (simplest) $N = 2$ case:

$$
\begin{aligned}
\ddot{x}_n = \frac{(-1)^{n+1}}{x_1 - x_2} \Big\{ &\, 2\,\dot{x}_1\,\dot{x}_2 \\
&+ \left[3\,\mathbf{i}\,\omega_1\,(\dot{x}_1 + \dot{x}_2) + 2\,(\omega_1)^2\,(x_1 + x_2)\right] x_1 \\
&- \left[3\,\mathbf{i}\,\omega_2\,(\dot{x}_1\,x_2 + x_1\,\dot{x}_2) + 2\,(\omega_2)^2\,x_1\,x_2\right] \\
&- c_1\,\left[\dot{x}_1 + \dot{x}_2 - \mathbf{i}\,\omega_1\,(x_1 + x_2)\right] x_1 \\
&- c_2\,\left[\dot{x}_1\,x_2 + x_1\,\dot{x}_2 - \mathbf{i}\,\omega_2\,x_1\,x_2\right] \Big\}, \\
&\qquad\qquad n = 1, 2.
\end{aligned}
\tag{4.176}
$$

And let us end this Section 4.5 with the following:

Remark 4.5.3. The two models considered in Section 4.5 are *Hamiltonian*, as implied by Remarks 2.4.5.2 and 4.2. ∎

4.6 Solvable *N*-Body Problems Identified via the Findings of Subsection 2.2.2

In Section 4.6 we introduce and discuss a class of *solvable* many-body problems of Newtonian type ("accelerations equal forces") obtained by using the identities reported in Subsection 2.2.2—in addition to the identities reported in Subsection 2.2.1 that were instrumental to obtain the results reported in the preceding sections of this Chapter 4. Some of these models are *multiply periodic, isochronous* or *asymptotically isochronous*; others display *scattering* phenomena. While these phenomenological features underline the potential appeal of these models, we anticipate that the interest of these findings is mainly in the additional vistas they offer on techniques to identify and investigate *new* classes of *solvable* many-body problems of Newtonian type.

Before displaying the equations of motion of the *solvable* class of models treated in Section 4.6, it is convenient to introduce some notation (for the convenience of the reader the following review of the notation used in Section 4.6 is more complete than strictly necessary, entailing some repetitions with respect to previous analogous notational presentations).

Notation 4.6.1. Unless otherwise indicated, above and hereafter N is an *arbitrary positive integer*, $N \geq 2$, indices such as n, m, ℓ, run over the *integers* from 1 to N, and superimposed arrows denote N-vectors: for instance the vector $\vec{c}$ has the N components c_m. A superimposed tilde denotes instead an *unordered* set of N numbers: for instance the notation $\tilde{z}$ denotes the *unordered* set of N numbers z_n. Uppercase **boldface** letters denote $N \times N$ matrices: for instance the matrix **M** features the N^2 elements M_{nm}. The numbers we use are generally assumed to be *complex*; except for those restricted to be *positive integers* (see above), which generally play the role of indices; and the "time" t, a variable generally restricted to *real* values. The *imaginary unit* is hereafter denoted as $\mathbf{i}$, implying of course $\mathbf{i}^2 = -1$. For quantities depending on the *real* independent variable t ("time"), superimposed dots indicate differentiation with respect to it: so, for instance, $\dot{z}_n(t) \equiv dz_n(t)/dt$, $\ddot{z}_n \equiv d^2 z_n/dt^2$; but often the t-dependence is not explicitly indicated whenever this is unlikely to cause misunderstandings. The Kronecker symbol δ_{nm} has the usual meaning: $\delta_{nm} = 1$ if $n = m$, $\delta_{nm} = 0$ if $n \neq m$; and we denote below as **I** the *unit* $N \times N$ matrix the elements of which are δ_{nm}. We adopt throughout the usual convention according to which a void sum vanishes and a void product equals unity: $\sum_{j=J}^{K} (\cdot)_j = 0$, $\prod_{j=J}^{K} (\cdot)_j = 1$ if $J > K$. Moreover we introduce the following convenient notations:

$$\sigma_m(\vec{z}) = \sum_{1 \leq s_1 < s_2 < \ldots < s_m \leq N} \left(z_{s_1} \cdot z_{s_2} \cdots z_{s_m} \right), \tag{4.177a}$$

$$\sigma_{n,m}(\vec{z}) = \delta_{1m} + \sum_{1 \leq s_1 < s_2 < \ldots < s_{m-1} \leq N \, ; \, s_j \neq n, \, j=1,\ldots m-1} \left(z_{s_1} \cdot z_{s_2} \cdots z_{s_{m-1}} \right), \tag{4.177b}$$

$$\sigma_{n_1 n_2, m}(\vec{z}) = \delta_{2m} + \sum_{\substack{1 \le s_1 < s_2 < \ldots < s_{m-2} \le N \, ; \\ s_j \ne n_1, \, s_j \ne n_2, \, j=1, \ldots m-2}} \left(z_{s_1} \cdot z_{s_2} \cdots z_{s_{m-2}} \right), \tag{4.177c}$$

$$\sigma_{n_1 n_2 n_3, m}(\vec{z}) = \delta_{3m} + \sum_{\substack{1 \le s_1 < s_2 < \ldots < s_{m-3} \le N \, ; \\ s_j \ne n_1, \, s_j \ne n_2, \, s_j \ne n_3, \, j=1, \ldots m-3}} \left(z_{s_1} \cdot z_{s_2} \cdots z_{s_{m-3}} \right), \tag{4.177d}$$

where the symbol $\sum_{1 \le s_1 < s_2 < \ldots < s_m \le N}$ denotes the sum from 1 to N over the m integer indices $s_1, s_2, \ldots, s_m$ with the restriction that $s_1 < s_2 < \ldots < s_m$. while the symbol $\sum_{1 \le s_1 < s_2 < \ldots < s_{m-1} \le N \, ; \, s_j \ne n, \, j=1, \ldots m-1}$ denotes the sum from 1 to N over the $m - 1$ indices $s_1, s_2, \ldots, s_{m-1}$ with the restriction $s_1 < s_2 < \ldots < s_{m-1}$ and moreover the requirement that *all these indices be different from* n; likewise for the symbols $\sum_{1 \le s_1 < s_2 < \ldots < s_{m-2} \le N \, ; \, s_j \ne n_1, \, s_j \ne n_2, \, j=1, \ldots, m-2}$ and $\sum_{1 \le s_1 < s_2 < \ldots < s_{m-3} \le N \, ; \, s_j \ne n_1, \, s_j \ne n_2, \, s_j \ne n_3, \, j=1, \ldots, m-3}$. Note that—according to the convention (see above) that a sum over an empty set of indices equals zero—these definitions imply $\sigma_{n,1}(\vec{z}) = 1$, they imply $\sigma_{n_1 n_2, 1}(\vec{z}) = 0$ and $\sigma_{n_1 n_2, 2}(\vec{z}) = 1$, and they imply $\sigma_{n_1 n_2 n_3, m}(\vec{z}) = 0$ for $m \le 2$ while $\sigma_{n_1 n_2 n_3, 3}(\vec{z}) = 1$. ∎

Remark 4.6.1. Note that the notation $\sigma_m(\tilde{z})$ (instead of $\sigma_m(\vec{z})$) is equally meaningful, since this quantity coincides with *symmetrical sums* of the N components z_m of the N-vector $\vec{z}$, hence it is independent of the ordering of the N elements z_n of the *unordered* set $\tilde{z}$, see (4.177a). The notations $\sigma_{n,m}(\vec{z})$, $\sigma_{n_1 n_2, m}(\vec{z})$, $\sigma_{n_1 n_2 n_3, m}(\vec{z})$, see (4.177), are instead ill-defined and therefore should not be used—except in the context of expressions which remain valid for *any* ordering of the N numbers z_n, i.e., for any assignments of the N different integer labels n (in the range $1 \le n \le N$) to the N elements of the *unordered* set $\tilde{z}$ provided that assignment is maintained throughout that expression (in which case the relevant expression amounts in fact to $N!$ equivalent formulas; assuming, as we generally do, that the N numbers z_n are *all different among themselves*). This remark is equally valid for any function $f(\tilde{z})$. ∎

The equations of motion of the specific $(2N)$-body model reported as example in Section 4.6 read as follows:

$$\ddot{z}_n = w_n, \tag{4.178a}$$

$$\ddot{w}_n = \sum_{\ell=1; \, \ell \ne n}^{N} \left(\frac{4 \, \dot{w}_n \, \dot{z}_\ell + 4 \, \dot{w}_\ell \, \dot{z}_n + 6 \, w_n \, w_\ell}{z_n - z_\ell} \right)$$

$$- 6 \sum_{\substack{\ell_1, \ell_2 = 1; \, \ell_1, \ell_2 \ne n; \ell_1 \ne \ell_2}}^{N} \left[\frac{w_n \, \dot{z}_{\ell_1} \, \dot{z}_{\ell_2} + 2 \, w_{\ell_1} \, \dot{z}_n \, \dot{z}_{\ell_2}}{\left(z_n - z_{\ell_1} \right) \left(z_n - z_{\ell_2} \right)} \right]$$

$$+ 4 \sum_{\substack{\ell_1, \ell_2, \ell_3 = 1; \, \ell_1, \ell_2, \ell_3 \ne n; \\ \ell_1 \ne \ell_2, \, \ell_2 \ne \ell_3, \, \ell_3 \ne \ell_1}}^{N} \left[\frac{\dot{z}_n \, \dot{z}_{\ell_1} \, \dot{z}_{\ell_2} \, \dot{z}_{\ell_3}}{\left(z_n - z_{\ell_1} \right) \left(z_n - z_{\ell_2} \right) \left(z_n - z_{\ell_3} \right)} \right]$$

$$- \left[\prod_{\ell=1,\ \ell\neq n}^{N} (z_n - z_\ell)^{-1} \right] \cdot$$

$$\cdot \sum_{m=1}^{N} \left[(\alpha_m \, \dddot{c}_m + \beta_m \, \ddot{c}_m + \gamma_m \, \dot{c}_m + \delta_m \, c_m) \ (z_n)^{N-m} \right]. \qquad (4.178b)$$

In these system of $2N$ equations of motion (4.178) the N coordinates $z_n(t)$ identify the positions in the *complex z*-plane of N particles (of type, say, "z"; identical among themselves) and the N coordinates $w_n(t)$ identify the positions in the *same complex* plane of N particles (of type, say, "w"; identical among themselves), the movements of which over time are determined by these equations of motion; the $4N$ quantities $c_m(t)$, $\dot{c}_m(t)$, $\ddot{c}_m(t)$, $\dddot{c}_m(t)$ in (4.178) are expressed in terms of these coordinates $z_n(t)$ and $w_n(t)$ and their time-derivatives by the following (*explicit!*: see (4.177)) formulas:

$$c_m = (-1)^m \, \sigma_m (\vec{z}) \equiv (-1)^m \, \sigma_m (\tilde{z}) \qquad (4.179a)$$

$$\dot{c}_m = (-1)^m \, \dot{\sigma}_m (\vec{z}) \equiv (-1)^m \, \sum_{n=1}^{N} \left[\sigma_{n,m} (\tilde{z}) \, \dot{z}_n \right], \qquad (4.179b)$$

$$\ddot{c}_m = (-1)^m \, \ddot{\sigma}_m (\vec{z}) \equiv (-1)^m \, \cdot$$

$$\cdot \left\{ \sum_{n=1}^{N} \left[\sigma_{n,m} (\tilde{z}) \, w_n \right] + \sum_{n_1,n_2=1, n_1\neq n_2}^{N} \left[\sigma_{n_1 n_2, m} (\tilde{z}) \, \dot{z}_{n_1} \, \dot{z}_{n_2} \right] \right\}, \qquad (4.179c)$$

$$\dddot{c}_m = (-1)^m \, \dddot{\sigma}_m (\vec{z}) \equiv (-1)^m \, \left\{ \sum_{n=1}^{N} \left[\sigma_{n,m} (\tilde{z}) \, \dot{w}_n \right] \right.$$

$$+ 3 \sum_{n_1,n_2=1, n_1\neq n_2}^{N} \left[\sigma_{n_1 n_2, m} (\tilde{z}) \, w_{n_1} \, \dot{z}_{n_2} \right]$$

$$\left. + \sum_{n_1,n_2,n_3=1, n_1\neq n_2\neq n_3}^{N} \left[\sigma_{n_1 n_2 n_3, m} (\tilde{z}) \, \dot{z}_{n_1} \, \dot{z}_{n_2} \, \dot{z}_{n_3} \right] \right\}, \qquad (4.179d)$$

where the indices n_1, n_2, n_3 in the last sum are distinct among themselves; while the parameters α_m, β_m, γ_m, δ_m in (4.178) are $4N$ *arbitrary complex* numbers, which may be conveniently related to the $8N$ *real* parameters $a_m^{(1)}$, $a_m^{(2)}$, $a_m^{(3)}$, $a_m^{(4)}$, $\omega_m^{(1)}$, $\omega_m^{(2)}$, $\omega_m^{(3)}$, $\omega_m^{(4)}$ (for their role see below equation (4.182)) by the following formulas:

$$\alpha_m = -a_m^{(1)} - a_m^{(2)} - a_m^{(3)} - a_m^{(4)} + \mathbf{i} \left[\omega_m^{(1)} + \omega_m^{(2)} + \omega_m^{(3)} + \omega_m^{(4)} \right], \qquad (4.180a)$$

$$\beta_m = -a_m^{(1)}a_m^{(2)} - a_m^{(1)}a_m^{(3)} - a_m^{(2)}a_m^{(3)} - a_m^{(1)}a_m^{(4)} - a_m^{(2)}a_m^{(4)} - a_m^{(3)}a_m^{(4)}$$
$$+ \omega_m^{(1)}\omega_m^{(2)} + \omega_m^{(1)}\omega_m^{(3)} + \omega_m^{(2)}\omega_m^{(3)} + \omega_m^{(1)}\omega_m^{(4)} + \omega_m^{(2)}\omega_m^{(4)} + \omega_m^{(3)}\omega_m^{(4)}$$
$$+ \mathbf{i}\Big[a_m^{(2)}\omega_m^{(1)} + a_m^{(3)}\omega_m^{(1)} + a_m^{(4)}\omega_m^{(1)} + a_m^{(1)}\omega_m^{(2)} + a_m^{(3)}\omega_m^{(2)} + a_m^{(4)}\omega_m^{(2)}$$
$$+ a_m^{(1)}\omega_m^{(3)} + a_m^{(2)}\omega_m^{(3)} + a_m^{(4)}\omega_m^{(3)} + a_m^{(1)}\omega_m^{(4)} + a_m^{(2)}\omega_m^{(4)} + a_m^{(3)}\omega_m^{(4)}\Big], \quad (4.180b)$$

$$\gamma_m = -a_m^{(1)}a_m^{(2)}a_m^{(3)} - a_m^{(1)}a_m^{(2)}a_m^{(4)} - a_m^{(1)}a_m^{(3)}a_m^{(4)} - a_m^{(2)}a_m^{(3)}a_m^{(4)}$$
$$+ a_m^{(1)}\omega_m^{(2)}\omega_m^{(3)} + a_m^{(1)}\omega_m^{(2)}\omega_m^{(4)} + a_m^{(1)}\omega_m^{(3)}\omega_m^{(4)} + a_m^{(2)}\omega_m^{(1)}\omega_m^{(3)}$$
$$+ a_m^{(2)}\omega_m^{(1)}\omega_m^{(4)} + a_m^{(2)}\omega_m^{(3)}\omega_m^{(4)} + a_m^{(3)}\omega_m^{(1)}\omega_m^{(2)} + a_m^{(3)}\omega_m^{(1)}\omega_m^{(4)}$$
$$+ a_m^{(3)}\omega_m^{(2)}\omega_m^{(4)} + a_m^{(4)}\omega_m^{(1)}\omega_m^{(2)} + a_m^{(4)}\omega_m^{(1)}\omega_m^{(3)} + a_m^{(4)}\omega_m^{(2)}\omega_m^{(3)}$$
$$+ \mathbf{i}\Big[a_m^{(2)}a_m^{(3)}\omega_m^{(1)} + a_m^{(2)}a_m^{(4)}\omega_m^{(1)} + a_m^{(3)}a_m^{(4)}\omega_m^{(1)} + a_m^{(1)}a_m^{(3)}\omega_m^{(2)}$$
$$+ a_m^{(1)}a_m^{(4)}\omega_m^{(2)} + a_m^{(3)}a_m^{(4)}\omega_m^{(2)} + a_m^{(1)}a_m^{(2)}\omega_m^{(3)} + a_m^{(1)}a_m^{(4)}\omega_m^{(3)}$$
$$+ a_m^{(2)}a_m^{(4)}\omega_m^{(3)} + a_m^{(1)}a_m^{(2)}\omega_m^{(4)} + a_m^{(1)}a_m^{(3)}\omega_m^{(4)} + a_m^{(2)}a_m^{(3)}\omega_m^{(4)}$$
$$- \omega_m^{(1)}\omega_m^{(2)}\omega_m^{(3)} - \omega_m^{(1)}\omega_m^{(2)}\omega_m^{(4)} - \omega_m^{(1)}\omega_m^{(3)}\omega_m^{(4)} - \omega_m^{(2)}\omega_m^{(3)}\omega_m^{(4)}\Big], \quad (4.180c)$$

$$\delta_m = -a_m^{(1)}a_m^{(2)}a_m^{(3)}a_m^{(4)} + a_m^{(3)}a_m^{(4)}\omega_m^{(1)}\omega_m^{(2)} + a_m^{(2)}a_m^{(4)}\omega_m^{(1)}\omega_m^{(3)}$$
$$+ a_m^{(2)}a_m^{(3)}\omega_m^{(1)}\omega_m^{(4)} + a_m^{(1)}a_m^{(4)}\omega_m^{(2)}\omega_m^{(3)} + a_m^{(1)}a_m^{(3)}\omega_m^{(2)}\omega_m^{(4)} + a_m^{(1)}a_m^{(2)}\omega_m^{(3)}\omega_m^{(4)}$$
$$- \omega_m^{(1)}\omega_m^{(2)}\omega_m^{(3)}\omega_m^{(4)} + \mathbf{i}\Big[a_m^{(2)}a_m^{(3)}a_m^{(4)}\omega_m^{(1)} + a_m^{(1)}a_m^{(3)}a_m^{(4)}\omega_m^{(2)}$$
$$+ a_m^{(1)}a_m^{(2)}a_m^{(4)}\omega_m^{(3)} + a_m^{(1)}a_m^{(2)}a_m^{(3)}\omega_m^{(4)} - a_m^{(4)}\omega_m^{(1)}\omega_m^{(2)}\omega_m^{(3)} - a_m^{(3)}\omega_m^{(1)}\omega_m^{(2)}\omega_m^{(4)}$$
$$- a_m^{(2)}\omega_m^{(1)}\omega_m^{(3)}\omega_m^{(4)} - a_m^{(1)}\omega_m^{(2)}\omega_m^{(3)}\omega_m^{(4)}\Big]. \quad (4.180d)$$

Remark 4.6.2. Let us note here that it is also possible to invert these equations, i.e., to write formulas expressing the $8N$ *real* parameters $a_m^{(1)}$, $a_m^{(2)}$, $a_m^{(3)}$, $a_m^{(4)}$, $\omega_m^{(1)}$, $\omega_m^{(2)}$, $\omega_m^{(3)}$, $\omega_m^{(4)}$ in terms of the $4N$ parameters α_m, β_m, γ_m, δ_m, but in view of the complicated nature of these expressions—essentially based on the solution of algebraic equations of fourth degree—this does not seem useful. ∎

The *solvable* character of this many-body problem is demonstrated by the following

Proposition 4.6.1. The *general solution* of the $(2N)$-body problem (4.178) with (4.179) is provided by the following prescription: the values of the coordinates $w_n(t)$ are given by (4.178a), while the values of the N coordinates $z_n(t)$ are the N zeros of the following monic time-dependent polynomial of degree N in the complex variable z,

$$P_N(z;t) = z^N + \sum_{m=1}^{N}\left[c_m(t)\, z^{N-m}\right], \quad (4.181)$$

the N *coefficients* $c_m(t)$ of which are provided—as shown below—by the following formulas:

$$c_m(t) = \sum_{k=1}^{4} \left\{ b_m^{(k)} \exp\left[\left(-a_m^{(k)} + \mathbf{i}\, \omega_m^{(k)} \right) t \right] \right\}. \tag{4.182}$$

Here the $8N$ parameters $a_m^{(k)}$ and $\omega_m^{(k)}$ are defined as above (see (4.180) and Remark 4.6.2), and the $4N$ *complex* numbers $b_m^{(k)}$ are $4N$ *a priori arbitrary* parameters; while in the context of the *initial-value problem* for this $(2N)$-body problem, (4.178) with (4.179), they are determined as the solutions, for every value of the parameter m, of the system of 4 *linear* algebraic equations

$$\sum_{k=1}^{4} \left[b_m^{(k)} \left(-a_m^{(k)} + \mathbf{i}\, \omega_m^{(k)} \right)^{s} \right] = \left. \frac{d^s c_m(t)}{dt^s} \right|_{t=0}, \quad s = 0, 1, 2, 3, \tag{4.183}$$

with, in the right-hand side, $c_m(0)$, $\dot{c}_m(0)$ expressed in terms of the initial data $z_n(0)$, $\dot{z}_n(0)$ by (4.179a) and (4.179b) (at $t = 0$), and $\ddot{c}_m(0)$, $\dddot{c}_m(0)$ expressed in terms of the initial data $z_n(0)$, $\dot{z}_n(0)$ and $w_n(0)$, $\dot{w}_n(0)$ by (4.179c) and (4.179d) (at $t = 0$). ■

Remark 4.6.3. Above and hereafter we assume for simplicity that the $4N$ complex numbers $\lambda_{m,k} = -a_m^{(k)} + \mathbf{i}\, \omega_m^{(k)}$ (see (4.182)) are *all different among themselves*; otherwise some appropriate limit should be taken in (4.182) and some of the statements made in the following Remark 4.6.4 would require additional clarifications. ■

Remark 4.6.4. The following properties of various subcases of the many-body problem characterized by the Newtonian equations of motion (4.178) are obviously implied by its *general solution*, as detailed above.

(i) If the $4N$ *real* parameters $a_m^{(k)}$ are *all nonnegative*, $a_m^{(k)} \geq 0$, then *all* solutions of this many-body problem are for *all future time* confined to a finite region—the dimensions of which depend on the initial data—of the *complex* plane; and in particular if the $4N$ *real* parameters $a_m^{(k)}$ are *all positive*, $a_m^{(k)} > 0$, *all* solutions of this many-body problem converge to the origin,

$$\lim_{t \to \infty} [z_n(t)] = 0, \quad \lim_{t \to \infty} [w_n(t)] = 0 ; \tag{4.184}$$

while if only *some* of the $4N$ *real* parameters $a_m^{(k)}$ are *positive* and *all* others *vanish*, then this many-body problem is *asymptotically multiply periodic*. If in addition the N *real* parameters $\omega_m^{(k)}$ corresponding to *vanishing* parameters $a_m^{(k)}$ are *all rational multiples* of a common (nonvanishing) *real* factor $\omega \neq 0$, i.e., if for *some* values of the indices m and k the parameters $a_m^{(k)}$ are positive, $a_m^{(k)} > 0$, while for *all other* values of the indices m and k

$$a_m^{(k)} = 0, \quad \omega_m^{(k)} = r_{mk}\, \omega, \tag{4.185}$$

with these parameters r_{mk} being *all rational* numbers (*positive* or *negative*, but *all different among themselves*), then this many-body system is *asymptotically isochronous* (for the meaning of this term see, if need be, [54] or Chapter 6 of [27]).

(ii) If the $4N$ *real* parameters $a_m^{(k)}$ all vanish, $a_m^{(k)} = 0$, and the $4N$ *real* parameters $\omega_m^{(k)}$ are *all multiples* r_{mk} of a *common* (*nonvanishing*) *real* factor $\omega \neq 0$, $\omega_m^{(k)} = r_{mk}\omega$ with the $4N$ parameters r_{mk} *all rational* numbers (*positive* or *negative* and *all different among themselves*), then this many-body system is *isochronous*: *all* its *generic* solutions are *completely periodic* with a period that is an *integer multiple* of the basic period $T = 2\pi / |\omega|$).

(iii) If *some* of the $4N$ *real* parameters $a_m^{(k)}$ are *negative*, then the solutions of this many-body problem will *not* be confined for *all* time, indeed one or more particles will travel out *far away* in the *remote future*. For a detailed analysis of such behaviors see Appendix G ("Asymptotic behavior of the zeros of a polynomial whose coefficients diverge exponentially") of the book [20]. ∎

Remark 4.6.5. Let us however recall that knowledge of the configuration of a many-body system at time t, with the (generally *complex*) values of its *coordinates* given as the *unordered* set of the *zeros* of a known time-dependent polynomial, does *not* allow to identify the specific coordinate that has evolved over time from the assignment of its specific initial position and velocity; this additional information can only be gained by following over time the evolution of the system, either by integrating numerically the equations of motion, or by identifying the configurations of the system by the technique described above at a sequence of time intervals sufficiently close to each other so as to guarantee the identification by *contiguity* of the trajectory of each particle (or at least of the specific particle under consideration). But these additional operations need not be performed with great accuracy, even when one wishes the final configuration—including the identity of each particle—to be known with much greater accuracy.

Likewise—in the case of systems that have been identified as *isochronous* because their solution is provided by the zeros of a time-dependent polynomial, which is itself periodic in time with period, say, T—an analogous procedure must be followed to ascertain whether the period of the time evolution of a specific particle is T, or νT (with ν a *positive integer*), due to the possibility of a T-periodic exchange of the correspondence between the zeros of the polynomial and the particle identities (see Section 2.3). ∎

Let us finally prove Proposition 4.6.1.

The key formulas to achieve this task are—in close analogy to previous developments in this book—the identities relating the time evolutions of the N *zeros* $z_n(t)$ of a time-dependent (monic) polynomial to those of the N *coefficients* $c_m(t)$ of the same polynomial (see, up to obvious notational changes—i.e. the replacement of $c_m(t)$ with $y_m(t)$ and of $z_n(t)$ with $x_n(t)$—the key formulas reported above in Subsections 2.2.1 and 2.2.2) as well as the relations (4.179) expressing the

N coefficients $c_m(t)$ of a time-dependent monic polynomial of degree *N* and their time-derivatives in terms of the *N zeros* $z_n(t)$ of the same polynomial and their time derivatives. With these formulas being complemented by the assumption that the *N* coefficients $c_m(t)$ evolve in time according to the following system of *N decoupled linear* ODEs,

$$\ddddot{c}_m = \alpha_m \, \dddot{c}_m + \beta_m \, \ddot{c}_m + \gamma_m \, \dot{c}_m + \delta_m. \tag{4.186}$$

Here the *4N* parameters α_m, β_m, γ_m, δ_m are the *4N complex* numbers introduced above, see (4.180). We assume that they imply the generic property that the characteristic fourth-degree algebraic equations associated to each of these *N* ODEs, reading

$$(\lambda_m)^4 - \left[\alpha_m \, (\lambda_m)^3 + \beta_m \, (\lambda_m)^2 + \gamma_m \, \lambda_m + \delta_m \right]$$

$$= \prod_{s=1}^{4} \left(\lambda_m - \lambda_{m,s} \right) = 0, \tag{4.187a}$$

has 4 distinct, generally *complex*, roots $\lambda_{m,s}$, $s = 1, 2, 3, 4$, which define the *8N real* parameters $a_m^{(1)}, a_m^{(2)}, a_m^{(3)}, a_m^{(4)}, \omega_m^{(1)}, \omega_m^{(2)}, \omega_m^{(3)}, \omega_m^{(4)}$ via the relations

$$\lambda_{m,s} = -a_m^{(s)} + \mathbf{i} \, \omega_m^{(s)}, \quad s = 1, 2, 3, 4, \tag{4.187b}$$

implying that the general solution of each of the ODEs (4.186) can be written as (4.182). Note that these developments are consistent with the fact that the relations (4.187a) with (4.187b) correspond just to the formulas (4.180), and that they imply that the *N* coordinates $z_n(t)$ evolve in time according to the equations of motions

$$\ddddot{z}_n = \sum_{\ell=1, \, \ell \neq n}^{N} \left(\frac{4 \, \dddot{z}_n \, \dot{z}_\ell + 4 \, \dddot{z}_\ell \, \dot{z}_n + 6 \, \ddot{z}_n \, \ddot{z}_\ell}{z_n - z_\ell} \right)$$

$$- 6 \sum_{\ell_1, \ell_2 = 1; \, \ell_1, \ell_2 \neq n; \, \ell_1 \neq \ell_2}^{N} \left[\frac{\ddot{z}_n \, \dot{z}_{\ell_1} \, \dot{z}_{\ell_2} + 2 \, \ddot{z}_{\ell_1} \, \dot{z}_n \, \dot{z}_{\ell_2}}{\left(z_n - z_{\ell_1} \right) \left(z_n - z_{\ell_2} \right)} \right]$$

$$+ 4 \sum_{\substack{\ell_1, \ell_2, \ell_3 = 1; \, \ell_1, \ell_2, \ell_3 \neq n; \\ \ell_1 \neq \ell_2, \, \ell_2 \neq \ell_3, \, \ell_3 \neq \ell_1}}^{N} \left[\frac{\dot{z}_n \, \dot{z}_{\ell_1} \, \dot{z}_{\ell_2} \, \dot{z}_{\ell_3}}{\left(z_n - z_{\ell_1} \right) \left(z_n - z_{\ell_2} \right) \left(z_n - z_{\ell_3} \right)} \right]$$

$$- \left[\prod_{\ell=1, \, \ell \neq n}^{N} \left(z_n - z_\ell \right)^{-1} \right] \cdot$$

$$\cdot \sum_{m=1}^{N} \left[\left(\alpha_m \, \dddot{c}_m + \beta_m \, \ddot{c}_m + \gamma_m \, \dot{c}_m + \delta_m \, c_m \right) \left(z_n \right)^{N-m} \right]. \tag{4.188}$$

In the last term the 4 quantities $\dddot{c}_m$, $\ddot{c}_m$, $\dot{c}_m$, and c_m must be expressed in terms of the dependent variables z_n and their time derivatives by the standard formulas expressing the *N coefficients*, and their time derivatives, of a time-dependent monic polynomial of degree N in terms of its N zeros and their time-derivatives:

$$c_m = (-1)^m \ \sigma_m \ (\vec{z}) \equiv (-1)^m \ \sigma_m \ (\tilde{z}) \ , \tag{4.189a}$$

$$\dot{c}_m = (-1)^m \ \dot{\sigma}_m \ (\vec{z}) \equiv (-1)^m \ \sum_{n=1}^{N} \left[\sigma_{n,m} \ (\tilde{z}) \ \dot{z}_n \right], \tag{4.189b}$$

$$\ddot{c}_m = (-1)^m \ \ddot{\sigma}_m \ (\vec{z}) \equiv (-1)^m \cdot$$

$$\cdot \left\{ \sum_{n=1}^{N} \left[\sigma_{n,m} \ (\tilde{z}) \ \ddot{z}_n \right] + \sum_{n_1,n_2=1,n_1 \neq n_2}^{N} \left[\sigma_{n_1 n_2,m} \ (\tilde{z}) \ \dot{z}_{n_1} \ \dot{z}_{n_2} \right] \right\} \tag{4.189c}$$

$$\dddot{c}_m = (-1)^m \ \dddot{\sigma}_m \ (\vec{z}) \equiv (-1)^m \ \left\{ \sum_{n=1}^{N} \left[\sigma_{n,m} \ (\tilde{z}) \ \dddot{z}_n \right] \right.$$

$$+ 3 \sum_{n_1,n_2=1,n_1 \neq n_2}^{N} \left[\sigma_{n_1 n_2,m} \ (\tilde{z}) \ \ddot{z}_{n_1} \ \dot{z}_{n_2} \right]$$

$$+ \sum_{n_1,n_2,n_3=1,n_1 \neq n_2 \neq n_3}^{N} \left[\sigma_{n_1 n_2 n_3,m} \ (\tilde{z}) \ \dot{z}_{n_1} \ \dot{z}_{n_2} \ \dot{z}_{n_3} \right] \right\}, \tag{4.189d}$$

where the indices n_1, n_2, n_3 in the last sum are distinct among themselves. Here the quantities $\sigma_m \ (\tilde{z})$, $\sigma_{n,m} \ (\tilde{z})$, $\sigma_{n_1 n_2,m}(\tilde{z})$, and $\sigma_{n_1 n_2 n_3,m} \ (\tilde{z})$ are defined by (4.177).

The proof of Proposition 4.6.1 is then accomplished by setting $w_n(t) = \dddot{z}_n(t)$ (see (4.178a)), implying the equivalence of (4.188) to (4.178), and likewise of (4.189) to (4.179).

Let us complete this Section 4.6 with the following

Remark 4.6.6. If in the (last term in the) right-hand side of (4.188) any one of the parameters α_m, β_m, γ_m, δ_m is independent of the index m, say $\beta_m = \beta$, then the corresponding term can be replaced by a simpler expression via the appropriate identity—see those reported in Subsections 2.2.1 and 2.2.2—implying, for instance,

$$\left[\prod_{\ell=1, \ \ell \neq n}^{N} (z_n - z_\ell)^{-1} \right] \sum_{m=1}^{N} \left[\beta \ \ddot{c}_m \ (z_n)^{N-m} \right] = \beta \left[\ddot{z}_n - \sum_{\ell=1, \ \ell \neq n}^{N} \left(\frac{2 \ \dot{z}_n \ \dot{z}_\ell}{z_n - z_\ell} \right) \right]. \ \blacksquare$$

$$\tag{4.190}$$

The alert reader will immediately understand the analogous simplifications implied by this Remark 4.6.6 for the equations of motion (4.178).

And let us end this Section 4.6 by displaying the equations of motion (4.178) with (4.179) in the simpler cases with $N = 2$ and $N = 3$.

Example 4.6.1. If $N = 2$ system (4.178) reduces to

$$\ddot{z}_n = w_n,$$
$$\ddot{w}_n = \frac{(-1)^{n+1}}{z_1 - z_2} \left[G(\vec{z}, \dot{\vec{z}}, \vec{w}, \dot{\vec{w}}) + z_n F_1(\vec{z}, \dot{\vec{z}}, \vec{w}, \dot{\vec{w}}) - F_2(\vec{z}, \dot{\vec{z}}, \vec{w}, \dot{\vec{w}}) \right],$$
$$n = 1, 2 \tag{4.191a}$$

where

$$G(\vec{z}, \dot{\vec{z}}, \vec{w}, \dot{\vec{w}}) = 4\,\dot{w}_1\,\dot{z}_2 + 4\,\dot{w}_2\,\dot{z}_1 + 6\,w_1\,w_2,$$
$$F_1(\vec{z}, \dot{\vec{z}}, \vec{w}, \dot{\vec{w}}) = \alpha_1\,(\dot{w}_1 + \dot{w}_2) + \beta_1\,(w_1 + w_2)$$
$$\qquad\qquad + \gamma_1\,(\dot{z}_1 + \dot{z}_2) + \delta_1\,(z_1 + z_2),$$
$$F_2(\vec{z}, \dot{\vec{z}}, \vec{w}, \dot{\vec{w}}) = \alpha_2\,(\dot{w}_1\,z_2 + 3\,w_1\,\dot{z}_2 + 3\,\dot{z}_1\,w_2 + z_1\,\dot{w}_2)$$
$$\qquad\qquad + \beta_2\,(w_1\,z_2 + 2\,\dot{z}_1\,\dot{z}_2 + z_1\,w_2) + \gamma_2\,(\dot{z}_1\,z_2 + z_1\,\dot{z}_2) + \delta_2\,z_1\,z_2.$$
■
$$\tag{4.191b}$$

Example 4.6.2. If $N = 3$ system (4.178) reads as follows:

$$\ddot{z}_n = w_n, \quad n = 1, 2, 3,$$
$$\ddot{w}_1 = \frac{4\dot{w}_1\dot{z}_2 + 4\dot{w}_2\dot{z}_1 + 6w_1w_2}{z_1 - z_2} + \frac{4\dot{w}_1\dot{z}_3 + 4\dot{w}_3\dot{z}_1 + 6w_1w_3}{z_1 - z_3}$$
$$\quad - \frac{1}{(z_1 - z_2)(z_1 - z_3)} \Big\{ 12\,[w_1\dot{z}_2\dot{z}_3 + \dot{z}_1(w_2\dot{z}_3 + w_3\dot{z}_2)]$$
$$\quad + (z_1)^2\,K_1(\vec{z}, \dot{\vec{z}}, \vec{w}, \dot{\vec{w}}) + z_1 K_2(\vec{z}, \dot{\vec{z}}, \vec{w}, \dot{\vec{w}}) + K_3(\vec{z}, \dot{\vec{z}}, \vec{w}, \dot{\vec{w}}) \Big\},$$
$$\ddot{w}_2 = \frac{4\dot{w}_2\dot{z}_1 + 4\dot{w}_1\dot{z}_2 + 6w_1w_2}{z_2 - z_1} + \frac{4\dot{w}_2\dot{z}_3 + 4\dot{w}_3\dot{z}_2 + 6w_2w_3}{z_2 - z_3}$$
$$\quad - \frac{1}{(z_2 - z_1)(z_2 - z_3)} \Big\{ 12\,[w_2\dot{z}_1\dot{z}_3 + \dot{z}_2(w_1\dot{z}_3 + w_3\dot{z}_1)]$$
$$\quad + (z_2)^2\,K_1(\vec{z}, \dot{\vec{z}}, \vec{w}, \dot{\vec{w}}) + z_2 K_2(\vec{z}, \dot{\vec{z}}, \vec{w}, \dot{\vec{w}}) + K_3(\vec{z}, \dot{\vec{z}}, \vec{w}, \dot{\vec{w}}) \Big\},$$
$$\ddot{w}_3 = \frac{4\dot{w}_3\dot{z}_1 + 4\dot{w}_1\dot{z}_3 + 6w_1w_3}{z_3 - z_1} + \frac{4\dot{w}_3\dot{z}_2 + 4\dot{w}_2\dot{z}_3 + 6w_2w_3}{z_3 - z_2}$$
$$\quad - \frac{1}{(z_3 - z_1)(z_3 - z_2)} \Big\{ 12\,[w_3\dot{z}_1\dot{z}_2 + \dot{z}_3(w_1\dot{z}_2 + w_2\dot{z}_1)]$$
$$\quad + (z_3)^2\,K_1(\vec{z}, \dot{\vec{z}}, \vec{w}, \dot{\vec{w}}) + z_3 K_2(\vec{z}, \dot{\vec{z}}, \vec{w}, \dot{\vec{w}}) + K_3(\vec{z}, \dot{\vec{z}}, \vec{w}, \dot{\vec{w}}) \Big\}, \tag{4.192a}$$

where

$$K_1(\vec{z}, \dot{\vec{z}}, \vec{w}, \dot{\vec{w}}) = -\big[\alpha_1 \,(\dot{w}_1 + \dot{w}_2 + \dot{w}_3) + \beta_1 \,(w_1 + w_2 + w_3)$$
$$+ \gamma_1 \,(\dot{z}_1 + \dot{z}_2 + \dot{z}_3) + \delta_1 \,(z_1 + z_2 + z_3)\big], \tag{4.192b}$$

$$K_2(\vec{z}, \dot{\vec{z}}, \vec{w}, \dot{\vec{w}}) = \alpha_2\big[\dot{w}_1(z_2 + z_3) + \dot{w}_2(z_1 + z_3) + \dot{w}_3(z_1 + z_2) + 2w_1(\dot{z}_2 + \dot{z}_3) + 2w_2(\dot{z}_1 + \dot{z}_3)$$
$$+ 2w_3(\dot{z}_1 + \dot{z}_2) + \dot{z}_1(w_2 + w_3) + \dot{z}_2(w_1 + w_3) + \dot{z}_3(w_1 + w_2)\big]$$
$$+ \beta_2\big[w_1(z_2 + z_3) + w_2(z_1 + z_3) + w_3(z_1 + z_2) + \dot{z}_1(\dot{z}_2 + \dot{z}_3) + \dot{z}_2(\dot{z}_1 + \dot{z}_3)$$
$$+ \dot{z}_3(\dot{z}_1 + \dot{z}_2)\big] + \gamma_2\big[\dot{z}_1(z_2 + z_3) + \dot{z}_2(z_1 + z_3) + \dot{z}_3(z_1 + z_2)\big]$$
$$+ \delta_2(z_1 z_2 + z_1 z_3 + z_2 z_3), \tag{4.192c}$$

$$K_3(\vec{z}, \dot{\vec{z}}, \vec{w}, \dot{\vec{w}}) = -\big\{\alpha_3\,[\dot{w}_1 z_2 z_3 + \dot{w}_2 z_1 z_3 + \dot{w}_3 z_1 z_2 + 2w_1(\dot{z}_2 z_3 + z_2 \dot{z}_3)$$
$$+ 2w_2(\dot{z}_1 z_3 + z_1 \dot{z}_3) + 2w_3(\dot{z}_1 z_2 + z_1 \dot{z}_2) + \dot{z}_1(w_2 z_3 + 2\dot{z}_2 \dot{z}_3 + z_2 w_3)$$
$$+ \dot{z}_2(w_1 z_3 + 2\dot{z}_1 \dot{z}_3 + z_1 w_3) + \dot{z}_3(w_1 z_2 + 2\dot{z}_1 \dot{z}_2 + z_1 w_2)]$$
$$+ \beta_3\,[w_1 z_2 z_3 + w_2 z_1 z_3 + w_3 z_1 z_2 + \dot{z}_1(\dot{z}_2 z_3 + z_2 \dot{z}_3) + \dot{z}_2(\dot{z}_1 z_3 + z_1 \dot{z}_3)$$
$$+ \dot{z}_3(\dot{z}_1 z_2 + z_1 \dot{z}_2)] + \gamma_3\,[\dot{z}_1 z_2 z_3 + z_1 \dot{z}_2 z_3 + z_1 z_2 \dot{z}_3] + \delta_3 z_1 z_2 z_3\big\}. \tag{4.192d}$$

■

4.7 Solvable Dynamical Systems (Sets of First-Order ODEs)

In the two subsections of Section 4.7, we report and tersely discuss two examples of *solvable* dynamical systems—i.e., sets of *first-order nonlinear* ODEs—the solutions of which can be achieved by *algebraic* operations and therefore display simple phenomenological behaviors that can be described rather comprehensively, see below. But clearly the main interest of these findings is in the vistas they open about the possibility to identify in analogous manners *many more* dynamical systems amenable to such *exact* treatments.

4.7.1 An Isochronous Dynamical System

The starting point of this example is the following set of *N decoupled* first-order ODEs satisfied by the *N* coefficients of the polynomial (2.3):

$$\dot{y}_m = \left(\frac{\mathbf{i}\, \rho_m \,\omega}{r_m - 1}\right) y_m + c_m \,(y_m)^{r_m}, \tag{4.193a}$$

where the *N* parameters c_m are *N arbitrary* (possibly *complex*) numbers, ω is an *arbitrary* (*nonvanishing*) *real* parameter to which we associate as usual the basic period

$$T = \frac{2\,\pi}{|\omega|}, \tag{4.193b}$$

and the $2N$ parameters r_m and ρ_m are *nonvanishing real rational* numbers (*arbitrary* except for the obvious conditions $r_m \neq 1$ and $\rho_m \neq 0$):

$$r_m = \frac{q_m}{k_m}, \quad \rho_m = \frac{\lambda_m}{\mu_m}, \qquad (4.193c)$$

with q_m and k_m coprime integers ($k_m \geq 1$, $k_m \neq q_m$, $q_m \neq 0$) and λ_m and μ_m coprime integers ($\mu_m \geq 1$, $\lambda_m \neq 0$).

It is then easily seen that the solution of the initial-value problem for these ODEs reads

$$y_m(t) = y_m(0) \left[\frac{\exp\left(-\mathbf{i}\,\rho_m\,\omega\,t\right) - b_m}{1 - b_m} \right]^{1/(1-r_m)}, \qquad (4.194a)$$

$$b_m = \left\{ 1 + \frac{\mathbf{i}\,\rho_m\,\omega\,\left[y_m(0)\right]^{1-r_m}}{(r_m - 1)\,c_m} \right\}^{-1}. \qquad (4.194b)$$

These formulas imply the following

Proposition 4.7.1.1. The solution of the initial-value problem for the dynamical system characterized by the system of N *coupled nonlinear* ODEs

$$\dot{x}_n = -\left[\prod_{\ell=1,\,\ell\neq n}^{N} (x_n - x_\ell) \right]^{-1} \sum_{m=1}^{N} \left\{ \left[\left(\frac{\mathbf{i}\,\rho_m\,\omega}{r_m - 1}\right) y_m + c_m\,(y_m)^{r_m} \right] (x_n)^{N-m} \right\},$$

$$(4.195)$$

is provided by the N *zeros* $x_n(t)$ of the polynomial (2.3) with *coefficients* $y_m(t)$, see (4.194). Both in the formulas (4.194) and in these equations of motion (4.195) the N *coefficients* $y_m(t)$ are given in terms of the N *zeros* $x_n(t)$ by the formulas (2.4) with $t = 0$ respectively $t = t$. ∎

The proof of this Proposition 4.7.1.1 is an immediate consequence of the identity (2.7) together with (4.193a) and (4.194).

Remark 4.7.1.1. It is evident from (4.194) that, considered as a function of the *real* variable t ("time"), $y_m(t)$ is *never singular* in the *nongeneric* case in which the initial datum $y_m(0)$ implies $|b_m| \neq 1$, i.e.

$$\left| 1 + \frac{\mathbf{i}\,\rho_m\,\omega\,\left[y_m(0)\right]^{1-r_m}}{(r_m - 1)\,c_m} \right| \neq 1; \qquad (4.196)$$

moreover it is *periodic*, for *generic* initial data, with period $T(\mu_m) = \mu_m T$ (see (4.193b) and (4.193c)),

$$y_m(t + T(\mu_m)) = y_m(t), \quad T = \frac{2\,\pi}{|\omega|}, \qquad (4.197)$$

if $1/(1 - r_m) = k_m/(k_m - q_m)$ is an *integer number* and, more generally, for *any* assignment of r_m if the initial datum $y_m(0)$ implies $|b_m| > 1$, i.e. if

$$\left| 1 + \frac{\mathbf{i}\, \rho_m\, \omega\, \left[y_m(0)\right]^{1-r_m}}{(r_m - 1)\, c_m} \right| < 1 \qquad (4.198a)$$

(since in this case the branch point of $y_m(t)$ falls *outside* of the circle—centered at the origin and of *unit* radius—traveled by the *complex* number $\exp\left(-\mathbf{i}\, \rho_m\, \omega\, t\right)$, see (4.194) as function of the *real* variable t; while it is also *periodic*, but with at most (see Remark 2.4.1.2) the *larger* period $M_m T$,

$$y_m\left(t + M_m\, T\right) = y_m(t), \quad M_m = |\text{MinimumCommonMultiple}\,[k_m, \mu_m]| \qquad (4.199a)$$

(see (4.193b) and (4.193c)) if the initial datum $y_m(0)$ implies that the *complementary* inequality to (4.198a) holds, i.e.

$$\left| 1 + \frac{\mathbf{i}\, \rho_m\, \omega\, \left[y_m(0)\right]^{1-r_m}}{(r_m - 1)\, c_m} \right| > 1. \ \blacksquare \qquad (4.199b)$$

This implies that the polynomial (2.3) with the N *coefficients* $y_m(t)$, see (4.194), is itself *periodic*—with a period that is an *integer* multiple of the basic period T hence *all generic* solutions of the dynamical system (4.195) are also *periodic* with some integer multiple of the basic period T (see Section 2.3), validating the fact that this dynamical system (4.195) is *isochronous*, as advertised in the title of this Subsection 4.7.1.

We complete this Subsection 4.7.1 by displaying the equations of motion (4.195) in the simplest case with $N = 2$:

$$\dot{x}_n = \frac{(-1)^n}{x_1 - x_2} \left\{ \left[-\left(\frac{\mathbf{i}\, \rho_1\, \omega}{r_1 - 1}\right)(x_1 + x_2) + (-1)^{r_1}\, c_1\, (x_1 + x_2)^{r_1} \right] x_n \right.$$
$$\left. + \left(\frac{\mathbf{i}\, \rho_2\, \omega}{r_2 - 1}\right) x_1\, x_2 + c_2\, (x_1\, x_2)^{r_2} \right\}, \quad n = 1, 2; \qquad (4.200a)$$

as well as its special case with $r_1 = r_2 = 2$ and $\rho_1 = \rho_2 = \rho$,

$$\dot{x}_n = \frac{(-1)^n}{x_1 - x_2} \left[-\mathbf{i}\, \rho\, \omega\, (x_n)^2 + c_1\, (x_1 + x_2)^2\, x_n + c_2\, (x_1\, x_2)^2 \right], \quad n = 1, 2. \qquad (4.200b)$$

4.7.2 Another Isochronous Dynamical System

Let us immediately report the main finding, via the following

Proposition 4.7.2.1. Let a dynamical system be defined by the following set of N *nonlinearly coupled first-order* ODEs characterizing the time-evolution of the N time-dependent variables $x_n(t)$:

$$
\dot{x}_n = \left[\prod_{\ell=1,\ \ell \neq n}^{N} (x_n - x_\ell)^{-1} \right] \left\{ \left[\mathbf{i}\, r\, \omega\, y_1 - c_1\, (y_1)^{(r-1)/r} \right]\ (x_n)^{N-1} \right.
$$
$$
\left. + \sum_{m=2}^{N} \left\{ \left[\mathbf{i}\, (r+m-1)\, \omega\, y_m - c_m\, y_{m-1} \right]\ (x_n)^{N-m} \right\} \right\}. \tag{4.201a}
$$

Here and below the N parameters c_m are N *arbitrarily assigned* (*nonvanishing*, possibly *complex*) numbers; the parameter ω is an *arbitrary nonvanishing real* number to which as usual the basic period $T = 2\pi / |\omega|$ is associated; the parameter r is an *arbitrarily assigned real rational* number,

$$
r = \frac{q}{k}, \tag{4.201b}
$$

with k and q two *coprime* integers ($k \geq 1$, $q \neq 0$); and the N functions $y_m(t)$ are supposed to be replaced by their expressions (2.4) in terms of the N coordinates $x_n(t)$ (so, they play the role of N *coefficients* of the monic polynomial (2.3) which features the N zeros $x_n(t)$). The *solution* of this dynamical system is then provided by the N zeros $x_n(t)$ of the polynomial (2.3) the *coefficients* $y_m(t)$ of which evolve as follows:

$$
y_m(t) = \exp\left[-\mathbf{i}\, (r+m-1)\, \omega\, t\right] \left\{ \sum_{\ell=0}^{m-2} \left(\left\{ \frac{\left[\exp(\mathbf{i}\,\omega\, t) - 1\right]^\ell}{(\mathbf{i}\,\omega)^\ell\, \ell!} \right\} \right. \right.
$$
$$
\cdot \left[C_{m,m-\ell+1}\, y_{m-\ell}(0) - C_{m,2}\, \beta_{m-\ell}\, y_1(0) \right] \bigg)
$$
$$
\left. + C_{m,2}\, \beta_m\, y_1(0) \left[1 + \alpha\, \frac{\exp(\mathbf{i}\,\omega\, t) - 1}{\mathbf{i}\,\omega} \right]^{r+m-1} \right\}, \quad m = 1, 2, \ldots, N,
$$
$$
\tag{4.202a}
$$

with $C_{m,\ell}$, β_m and α defined as follows:

$$
C_{m,\ell} = \prod_{s=\ell}^{m} (c_s), \quad \beta_m = \alpha^{1-m} \prod_{j=1}^{m-1} \left[(r+j)^{-1} \right], \quad \alpha = \frac{c_1}{r} \left[y_1(0) \right]^{-1/r}, \tag{4.202b}
$$

and the N quantities $y_m(0)$ defined in terms of the N initial data $x_n(0)$ by the formulas (2.4) at $t = 0$. ∎

Remark 4.7.2.1. If r is a *nonnegative integer* (i.e. if $k = 1$ and $q \geq 0$, see (4.201b)), for *arbitrary* initial data $y_m(0)$ these N functions $y_m(t)$, see (4.202), are *all nonsingular* and *periodic* with the basic period $T = 2\pi / |\omega|$,

$$y_m(t + T) = y_m(t). \tag{4.203}$$

Otherwise they might become *singular* for values of t such that $\exp(\mathbf{i}\,\omega\,t) = 1 - \mathbf{i}\,\omega/\alpha$, but this can only happen for *nongeneric* values of the initial datum $y_1(0)$ such that there hold the equality $|1 - \mathbf{i}\,\omega/\alpha| = 1$ namely

$$\left| 1 - \left(\frac{\mathbf{i}\,\omega\,r}{c_1} \right) \left[y_1(0) \right]^{1/r} \right| = 1. \tag{4.204a}$$

If there holds instead the *inequality*

$$\left| 1 - \left(\frac{\mathbf{i}\,\omega\,r}{c_1} \right) \left[y_1(0) \right]^{1/r} \right| > 1, \tag{4.204b}$$

then *all* the functions $y_m(t)$ are (*nonsingular* and) *periodic* with the basic period $T = 2\pi / |\omega|$. While if there holds the *complementary inequality*

$$\left| 1 - \left(\frac{\mathbf{i}\,\omega\,r}{c_1} \right) \left[y_1(0) \right]^{1/r} \right| < 1, \tag{4.204c}$$

then the corresponding functions $y_m(t)$ are still (*nonsingular* and) *periodic*, but with the larger period $kT = 2\pi k / |\omega|$,

$$y_m(t + k\,T) = y_m(t), \tag{4.205}$$

see (4.201b). ∎

Remark 4.7.2.2. The dynamical system (4.201a) is *isochronous*, *all* its *generic* (hence *nonsingular*) solutions being completely *periodic* with at most period

$$\tilde{T} = \nu_{\text{Max}}(N)\,\tilde{M}\,T, \quad \tilde{M} = \underset{m=1,2,\dots,N}{\text{MinimumCommonMultiple}}\,[k_m], \tag{4.206a}$$

$$x_n(t + \tilde{T}) = x_n(t), \quad n = 1, ,2, \dots, N\,; \tag{4.206b}$$

while many of them are *completely periodic* with a period νT with ν an *integer* in the range $1 \leq \nu \leq \nu_{\text{Max}}(N)$ (for the significance of the integers ν and $\nu_{\text{Max}}(N)$ see Section 2.3). ∎

We trust the reader to consider this Remark 4.7.2.2 obvious on the basis of the developments reported above: see in particular Section 2.3.

Let us conclude this Subsection 4.7.2 by proving Proposition 4.7.2.1.

Our starting point is the rather trivial dynamical system characterized by the following simple system of N Ordinary Differential Equations (ODEs) (here and below appended primes denote differentiations of the N, generally *complex*, dependent variables $\gamma_m(\tau)$ with respect to the, also *complex*—see below—independent variable τ):

$$\gamma_1'(\tau) = c_1 \left[\gamma_1(\tau) \right]^{(r-1)/r}, \tag{4.207a}$$

$$\gamma_m'(\tau) = c_m \, \gamma_{m-1}(\tau), \quad m = 2, \dots, N. \tag{4.207b}$$

As the reader may easily verify, the solution of the initial-value problem of this dynamical system reads as follows (we assume $\gamma_1(0) \neq 0$ if $r < 1$, see (4.207a)):

$$\gamma_m(\tau) = \sum_{\ell=0}^{m-2} \left\{ \frac{\tau^\ell}{\ell!} \left[C_{m,m-\ell+1} \, \gamma_{m-\ell}(0) - C_{m,2} \, \beta_{m-\ell} \, \gamma_1(0) \right] \right\}$$
$$+ \, C_{m,2} \, \beta_m \, \gamma_1(0) \, (1 + \alpha \, \tau)^{r+m-1}, \quad m = 1, 2, \dots, N, \tag{4.208}$$

with $C_{m,\ell}$, β_m and α defined as above (see (4.202b)).

A, perhaps more interesting, set of dynamical systems (now evolving in the *real* time t) obtains then via the position

$$\tau \equiv \tau(t) = \frac{\exp(\mathbf{i} \, \omega \, t) - 1}{\mathbf{i} \, \omega}, \tag{4.209a}$$

and

$$y_m(t) = \exp\left[-\mathbf{i} \, (r + m - 1) \, \omega \, t\right] \gamma_m(\tau(t)), \quad m = 1, 2, \dots, N, \tag{4.209b}$$

making thereby contact with the notation and results reported above. It is indeed then easily seen that the time-dependent variables $y_m(t)$ evolve according to the *autonomous* dynamical system characterized by the ODEs

$$\dot{y}_1(t) = -\mathbf{i} \, r \, \omega \, y_1(t) + c_1 \left[y_1(t) \right]^{(r-1)/r}, \tag{4.210a}$$

$$\dot{y}_m(t) = -\mathbf{i} \, (r + m - 1) \, \omega \, y_m(t) + c_m \, y_{m-1}(t), \quad m = 2, \dots, N, \tag{4.210b}$$

and that the solution of this dynamical system is provided just by the formula (4.202). While the insertion of the ODEs (4.210) in the right-hand side of the identity (2.7) yields just the equations of motion (4.201a). Proposition 4.7.2.1 is thereby proven.

Let us end this Subsection 4.7.2 by displaying the equations of motion (4.201a) in the simplest, $N = 2$, case:

$$\dot{x}_n = \frac{(-1)^{n+1}}{x_1 - x_2} \left[\left\{ -\mathbf{i} \, r \, \omega \, (x_1 + x_2) - c_1 \, \left[-(x_1 + x_2) \right]^{(r-1)/r} \right\} x_n \right.$$
$$\left. + \mathbf{i} \, (r + 1) \, \omega \, x_1 \, x_2 + c_2 \, (x_1 + x_2) \right], \quad n = 1, 2. \tag{4.211}$$

4.8 How to Manufacture Many More Solvable Dynamical Systems

In the preceding sections of this Chapter 4 quite a few dynamical systems have been identified and discussed that are *solvable* by algebraic operations. The basic idea to identify such models is to consider a *time-dependent* monic polynomial of degree N, see (2.3), the N coefficients $y_m \equiv y_m(t)$ of which evolve in a *solvable* manner and to then focus on the corresponding evolution of the N *zeros* $x_n \equiv x_n(t)$ of that polynomial. Above we generally—albeit not exclusively—restricted consideration to evolutions of the N coefficients $y_m(t)$ characterized by *analogous* ODEs— or possibly systems of ODEs—so that the relevant ODEs characterizing the time evolution of different coefficients $y_m(t)$ only differed due to different assignments of the *parameters* featured by these ODEs. The lifting of this restriction—which is by no means a necessary requirement for our approach—opens an immense range of possibilities, by allowing a much larger choice for the evolution of each of the N coefficients $y_m(t)$, and leading to a correspondingly immense set of *solvable* dynamical systems for the N zeros $x_n(t)$. The exploration of this universe is left to whoever will eventually find some motivation to do so—perhaps by identifying specific examples of theoretical or applicative interest. In Section 4.8 we merely limit our treatment to a quite simple example (but note that also in previous subsections of this Section 4.4 we occasionally took advantage of this freedom).

Let us therefore focus on the simplest case with $N = 2$, assuming that the two *coefficients* $y_1(t)$ and $y_2(t)$ of the polynomial

$$p_2(z;t) = z^2 + y_1(t) z + y_2 = [z - x_1(t)] \ [z - x_2(t)] \qquad (4.212)$$

evolve, say, as follows:

$$\ddot{y}_1 = \mathbf{i} \, \omega \, \dot{y}_1 + r_1 \, \omega^2 \, y_1 + \frac{(\dot{y}_1)^2}{y_1}, \qquad (4.213a)$$

$$\ddot{y}_2 = - (r_2 \, \omega)^2 \, y_2, \qquad (4.214a)$$

with ω an *arbitrary nonvanishing real* number to which we associate as usual the basic period $T = 2\pi / |\omega|$, and r_n two *rational* numbers,

$$r_n = \frac{q_n}{k_n}, \qquad (4.215)$$

with q_n and k_n two *coprime integers* ($q_n \neq 0, k_n \geq 1; n = 1, 2$). This clearly implies (see the case with $c = 1$ in Subsection 2.4.1, in particular equations (2.25) and (2.24b))

$$y_1(t) = y_1(0) \ \exp(\mathbf{i} \, r_1 \, \omega \, t) \ \exp\{a \ [\exp(\mathbf{i} \, \omega \, t) - 1]\},$$
$$a = \frac{\dot{y}_1(0)}{\mathbf{i} \, \omega \, y_1(0)} - r_1, \qquad (4.216a)$$

and

$$y_2(t) = y_2(0) \, \cos{(r_2 \, \omega \, t)} + \frac{\dot{y}_2(0) \, \sin{(r_2 \, \omega \, t)}}{r_2 \, \omega}. \qquad (4.216b)$$

Hence $y_1(t)$ is *periodic* with period $k_1 T$ and $y_2(t)$ is periodic with period $k_2 T$ (see (4.215)); therefore the polynomial $p_2 (z; t)$, see (4.212), is *periodic* with period KT where $K = \text{MinimumCommonMultiple}[k_1, k_2]$.

The two *zeros* $x_1(t)$, $x_2(t)$ of this polynomial $p_2 (z; t)$ are then given by the formula

$$x_n(t) = \frac{1}{2} \left[-y_1(t) + (-1)^{n+1} \left\{ [y_1(t)]^2 - 4 \, y_2(t) \right\}^{1/2} \right], \quad n = 1, 2 ; \quad (4.217)$$

they are clearly *periodic* functions of time with period KT or possibly $2KT$ (see Section 2.3).

As implied by our approach (see Chapter 2, in particular Section 2.2) this formula—with $y_1(t)$ and $y_2(t)$ expressed by (4.216) with $y_1(0) = -[x_1(0) + x_2(0)]$ and $y_2(0) = x_1(0)x_2(0)$—provides therefore the solution of the initial-value problem of the following system of 2 nonlinear Newtonian equations of motions for the 2 dependent variables $x_n(t)$:

$$\ddot{x}_n = \frac{(-1)^{n+1}}{x_1 - x_2} \left\{ 2 \, \dot{x}_1 \, \dot{x}_2 \right.$$
$$+ \left[\mathbf{i} \, \omega \, (\dot{x}_1 + \dot{x}_2) + r_1 \, \omega^2 \, (x_1 + x_2) + \frac{(\dot{x}_1 + \dot{x}_2)^2}{x_1 + x_2} \right] x_n$$
$$\left. + (r_2 \, \omega)^2 \, x_1 \, x_2 \right\}, \quad n = 1, 2. \qquad (4.218)$$

And it is a matter of trivial algebra—via the assignment

$$X(t) = x_1(t) + x_2(t), \quad x(t) = x_1(t) - x_2(t) \qquad (4.219)$$

—to obtain for the 2 new dependent variables $X(t)$ and $x(t)$ the following alternative system of *nonlinear Newtonian* ODEs:

$$\ddot{X} = \mathbf{i} \, \omega \, \dot{X} + r_1 \, \omega^2 \, X + \frac{\dot{X}^2}{X}, \qquad (4.220a)$$

$$\ddot{x} = -\frac{1}{2} \, (r_2 \, \omega)^2 \, x - \frac{\dot{x}^2}{x}$$
$$+ x^{-1} \left\{ [r_1 + (r_2)^2 /2] \, \omega^2 \, X^2 + \mathbf{i} \, \omega \, X \dot{X} + \dot{X}^2 \right\}. \qquad (4.220b)$$

The solution of the initial-value problem for this system of 2 Newtonian equations of motion—describing the motion of the 2 coordinates $X(t)$ and $x(t)$ in the complex plane—reads

$$X(t) = X(0) \, \exp\left(\mathbf{i} \, r_1 \, \omega \, t\right) \, \exp\left\{\rho_1 \left[\exp\left(\mathbf{i} \, \omega \, t\right) - 1\right]\right\},$$

$$\rho_1 = \frac{\dot{X}(0)}{\mathbf{i} \, \omega \, X(0)} - r_1, \tag{4.221a}$$

$$x(t) = \left\{ [X(t)]^2 + \left[x^2(0) - X^2(0)\right] \, \cos\left(r_2 \, \omega \, t\right) \right.$$

$$\left. + \frac{2 \left[x(0) \, \dot{x}(0) - X(0) \, \dot{X}(0)\right] \, \sin\left(r_2 \, \omega \, t\right)}{r_2 \, \omega} \right\}^{1/2}. \tag{4.221b}$$

And note that the first of these 2 evolutions, see (4.221a), is *nonsingular* for *any* assignment of the initial datum $X(0)$, and it is easily seen that the second, see (4.221b), is as well *nonsingular* for an *open* set of initial data: it is clearly sufficient that $|X(0)|^2$ be adequately *large,* and both $\left|x^2(0) - X^2(0)\right|$ and $\left|x(0) \, \dot{x}(0) - X(0) \, \dot{X}(0)\right|$ be adequately *small* (the alert reader will have no difficulty to make these conditions more precise). And it is evident that for this *open* set of *initial* data $X(t)$ is *periodic* with period $k_1 T$ and $x(t)$ is *periodic* with period KT.

The equations of motion (4.220) can be reformulated as equations of motion for 2 *real* 2-vectors $\vec{X}(t)$ and $\vec{x}(t)$ identifying 2 point-particles moving in the *real* Cartesian plane (instead of the 2 *complex* points $X(t)$ and $x(t)$ moving in the *complex* plane) via the standard formulas reported in the Appendix; there obtain in this manner the equations of motion (1.2) of Example 1.2 in Chapter 1.

4.N Notes on Chapter 4

As already mentioned in Chapter 1, the main tool to identify and investigate the *solvable/integrable* many-body models discussed in Chapter 4 are the relations among the time-evolutions of the *zeros* and the *coefficients* of a monic time-dependent polynomial such as (2.3). This approach was introduced a few decades ago in [16], but the technique to implement it was then essentially restricted to the consideration of the time-evolution of the *zeros* of a (monic) polynomial the *coefficients* of which evolve according to a (rather special) system of *linear* ODEs. The elimination of this restriction to *linear* ODEs for the coefficients was recently made possible by the key formulas reported in Chapter 2, opening thereby the way to the identification of many more *solvable/integrable* N-body models, as described in Chapter 4 and demonstrated by the examples reported there. Below we identify—for the convenience of the reader interested to pursue the investigation of some of these models—the references (if any) where each of the N-body models presented above were introduced and discussed. Other models—not reported in this book—can be found in the recent literature referred to in this Section 4.N.

The goldfish N-body problem discussed in Section 4.1 was identified as solvable in [16]. Its name "goldfish" was proposed in [21] as recognition for the neatness of its formulation and solution, with reference to the following quotation by Vladimir E. Zakharov: "A mathematician, using the dressing method to find a new integrable system, could be compared with a fisherman, plunging his net into the sea. He does not know what a fish he will pull out. He hopes to catch a goldfish, of course. But too often his catch is something that could not be used for any known to him purpose. He invents more and more sophisticated nets and equipments and plunges all that deeper and deeper. As a result he pulls on the shore after a hard work more and more strange creatures. He should not despair, nevertheless. The strange creatures may be interesting enough if you are not too pragmatic. And who knows how deep in the sea do goldfishes live?" [93].

Hamiltonians yielding the goldfish equations of motion (4.84c) were introduced in [49]; for a more complete recent treatment of the Hamiltonian structure associated to the goldfish model see [81].

The following (probably incomplete) list identifies papers in which various properties and variants of the goldfish model (see Section 4.1) have been treated: [20], [23], [25], [85], [89], [91], [78], [76], [55], [10], [58], [59], [11], [77], [53], [27], [1], [56], [28], [30], [75], [32], [33], [70], [2], [79]. To this list should be added the recent papers [34], [4], [5], [14], [37]-[42] which treat the other models—all of them "of goldfish type"—reported in Chapter 4, see above and below.

An interesting *physical* application of the goldfish model itself (in its simplest *nonisochronous* version, see (4.83a) with $\omega = 0$) is identified in the paper [83] (which is unfortunately marred by several misprints): see equation (6) there.

The main N-body problem treated in Section 4.2 was introduced and discussed in [42]. For the N-body models reported in the 3 Subsections 4.2.1,4.2.2,4.2.3 see [34], and for the N-body model reported in Subsection 4.2.4 see [40]; but note the nontrivial extension of the treatment presented herein for the *isochronous* variants.

For the N-body model reported in Subsection 4.2.5 see the treatment (including the graphical display of some solutions) in [4]. Let us recall in connection with this N-body model that the N-body problem characterized by the equations of motion (4.122a), hence by the Hamiltonian (4.122b), was introduced and solved in the *quantal* context in 1971 (see [15]), and solved in the *classical* context by Jürgen Moser in 1975 [84]; it is therefore generally known as the Calogero-Moser N-body model (this justifies the title of Subsection 4.2.5). Its *explicit* solution in the classical case—as reported above, see (4.123)—is due to M. A. Olshanetsky and A. M. Perelomov [86] (see also [20] and references therein).

For the N-body model reported in Subsection 4.2.6, see [37].

For the N-body model reported in Subsections 4.3.1, see Example 6 in [34] (but note a trivial misprint there in equations (13a,b)); for that treated in Subsection 4.3.2 see [38] (but note that the *isochronous* variant reported herein is more general than that treated there, to which it reduces for $r_m = mr$ with r a *real rational* number and $\rho_m = 1$).

The results reported in Subsection 4.3.5 and in Section 4.4 have not been previously published.

Of the two *solvable* Newtonian N-body models treated in Section 4.5, the first is a generalization of the third case treated in [39] (to which it reduces for the special assignment $r_m = mr$ and $\rho_m = 1$), and the second has not been previously published.

For the results reported in Section 4.6 see [3], where several graphs are also displayed of the solutions of two examples of the N-body problem (4.178) with $N = 2$ respectively $N = 3$.

The results reported in Subsection 4.7.1 have not been previously published; for those reported in Subsection 4.7.2 see [41].

The simple findings reported in Section 4.8 have not been previously published.

5

Solvable Systems of Nonlinear Partial Differential Equations (PDEs)

Up to this point we focused in this book on *dynamical systems*: sets of *N-coupled nonlinear ODEs* characterizing the time-evolution in the *complex x*-plane of *N* points the positions of which are identified by *N* coordinates $x_n(t)$ being the *dependent* variables of the *dynamical system*. The models we investigated are *algebraically solvable* by interpreting these coordinates $x_n(t)$ as the *N zeros* of a monic polynomial of degree *N*, the *N coefficients* of which are themselves evolving in time according to systems of *N algebraically solvable* ODEs. In particular, we mainly looked at such systems that are interpretable as *N-body problems* characterized by *Newtonian equations of motion*: "accelerations equal forces", with the forces appropriately defined in terms of the positions and velocities of the *N* moving particles. The identification of such models was essentially facilitated by convenient *identities* relating the time evolution of the *N zeros* of a time-dependent monic polynomial of degree *N* to the time evolution of its *N coefficients*: see Chapter 2.

In Chapter 5 we show how the same kind of approach allows to identify and investigate *new* systems of *nonlinearly coupled partial differential equations* (PDEs) of evolution *amenable to exact treatments*. As in the case of the findings reported in the previous chapters of this book, the main relevance of the findings reported in Chapter 5 comes from the possibility they offer to identify and investigate *new* systems of PDEs *amenable to exact treatments*, rather than from the importance of the specific examples we use below to demonstrate this technique. Indeed we limit our presentation below to just two examples of evolution PDEs: one "C-integrable" and one "S-integrable".

Remark 5.1. Let us recall the significance of this terminology [17], now rather largely used. An evolution equation is called *C-integrable* if it can be reduced by a—generally explicit—*Change* of (dependent and independent) variables to an equation that is itself *solvable* by elementary techniques (for instance it is *autonomous* and *linear*, hence *solvable* by a standard *integral transform* such as Fourier). Likewise, an evolution equation is called *S-integrable* if it can be *solved* via the so-called *Spectral* (or *Scattering*) transform: see, for instance, [46]. Consistently with the point of view on which this book is based, we also call *C-integrable* respectively *S-integrable* any system of evolution PDEs that can be transformed by *algebraic* operations—such as the transition from the *zeros* to the *coefficients* of a polynomial—to a *C-integrable* respectively a *S-integrable* evolution PDE. ∎

In particular, below we identify as examples of a *new* system of *C-integrable*, respectively *S-integrable*, nonlinearly coupled PDEs in $1 + 1$ (*real*) variables (space x and time t) those characterizing the evolution of the N zeros of a monic polynomial—of degree N in the (*complex*) variable z and depending on the two *real* variables x and t—featuring N coefficients each of which evolves according to a Burgers PDE (perhaps the most elementary *nonlinear C-integrable* PDE in $1 + 1$ dimensions), respectively as a Korteweg-de Vries (KdV) PDE (perhaps the most famous of the *nonlinear S-integrable* PDEs in $1+1$ dimensions, since the discovery [74] half a century ago of its integrability via the *Spectral* transform opened the way to a major development in pure and applied mathematics).

Notation 5.1. In Chapter 5 we use as essential tools the following *monic* polynomials of *arbitrary* degree N ($N \geq 2$) in the *complex* variable z,

$$P_N\big(z; \vec{\varphi}(x,t), \tilde{\psi}(x,t)\big) = z^N + \sum_{m=1}^{N} \left[\varphi_m(x,t)\, z^{N-m} \right] = \prod_{n=1}^{N} \left[z - \psi_n(x,t) \right],$$
$$(5.222)$$

which are analogous to those employed in the preceding chapters (see (2.3)), except for the fact that these polynomials depend now on a ("space") variable x in addition to the time t, causing an appropriate change of notation for the N coefficients $\varphi_m(x,t)$ and the N zeros $\psi_n(x,t)$ of these polynomials. The rest of the notation is, we trust, self-evident. Indices such as n, m again run throughout from 1 to N, the N-vector $\vec{\varphi}(x,t)$ has the N coefficients $\varphi_m(x,t)$ of the polynomial (5.222) as its N components, $\tilde{\psi}(x,t)$ denotes the *unordered* set of the N zeros $\psi_n(x,t)$ of the polynomial (5.222), and we generally assume all these dependent variables to be *complex* (this does not exclude that they might be *real*, see indeed the examples below). We instead assume the independent variables x ("space") and t ("time") to be *real* numbers (and we omit their explicit display when this can be done without causing misunderstandings); and we indicate partial differentiations with respects to these variables by appending them as subscripts preceded by commas, so, for instance, $\varphi_{m,t} \equiv \partial \varphi_m(x,t)/\partial t$, $\psi_{n,xx} \equiv \partial^2 \psi_n(x,t)/\partial x^2$. We generally focus on *generic* polynomials the *coefficients* and *zeros* of which are *generic complex* numbers, and which in particular feature *zeros* all *different among themselves*, $\psi_n(x,t) \neq \psi_m(x,t)$ if $n \neq m$ for all values of x and t. And let us reemphasize that the notation $P_N\big(z; \vec{\varphi}, \tilde{\psi}\big)$, see (5.222), is somewhat redundant, since this monic polynomial of degree N in z can be identified by assigning *either* its N coefficients $\varphi_m(x,t)$ (i.e., the N components of the N-vector $\vec{\varphi}(x,t)$) *or* its N zeros $\psi_n(x,t)$ (i.e., the N elements of the N-dimensional *unordered* set $\tilde{\psi}(x,t)$); indeed the N coefficients $\varphi_m(x,t)$ can be expressed in terms of the N zeros $\psi_n(x,t)$ via the standard formula (see (2.4))

$$\varphi_m = (-1)^m \sum_{1 \le n_1 < n_2 < \cdots < n_m \le N} \left(\psi_{n_1} \cdot \psi_{n_2} \cdots \psi_{n_m} \right), \tag{5.223a}$$

implying (see (2.5))

$$\varphi_{1,x} = -\sum_{n=1}^{N} \left(\psi_{n,x} \right), \tag{5.223b}$$

$$\varphi_{m,x} = (-1)^m \sum_{s=1}^{m} \left[\psi_{n_s,x} \cdot \right.$$

$$\left. \cdot \sum_{1 \le n_1 < n_2 < \cdots n_{s-1} < n_{s+1} < n_m \le N} \left(\psi_{n_1} \cdot \psi_{n_2} \cdots \psi_{n_{s-1}} \psi_{n_{s+1}} \cdots \psi_{n_m} \right) \right],$$

$$m = 2, \ldots, N. \tag{5.223c}$$

On the other hand, while the assignment of the *N coefficients* $\varphi_m(x,t)$ determines *uniquely*, up to permutations, the *N zeros* $\psi_n(x,t)$, *explicit* formulas in terms of *elementary* functions (essentially *radicals*) expressing the *zeros* of a polynomial of degree *N* in terms of its *coefficients* are generally *only* available for $N \le 4$. ■

In the following two Sections 5.1 respectively 5.2 we report 2 examples of *C-integrable* respectively *S-integrable* systems of PDEs in $1 + 1$ dimensions, and in Section 5.3 we prove the findings reported in Sections 5.1 and 5.2. And let us end this presentation of the contents of this Chapter 5 by re-emphasizing that the approach described in it provides the possibility to identify a large universe of new *integrable/solvable* systems of nonlinear PDEs: the two PDEs specifically discussed below are merely *examples* of the vistas opened by this methodology to identify new *integrable/solvable* systems of *nonlinear* PDEs (this last observation is reinforced in Section 5.4).

5.1 A C-Integrable System of Nonlinear PDEs

Proposition 5.1.1. The following system of *N* coupled nonlinear PDEs in $1 + 1$ variables is *C-integrable*:

$$\psi_{n,t} + \psi_{n,xx} = \sum_{\ell=1, \ell \ne n}^{N} \left(\frac{2 \, \psi_{n,x} \, \psi_{\ell,x}}{\psi_n - \psi_\ell} \right)$$

$$- \left[\prod_{\ell=1, \ell \ne n}^{N} \left(\psi_n - \psi_\ell \right) \right]^{-1} \sum_{m=1}^{N} \left[a_m \, \varphi_{m,x} \, \varphi_m \, \left(\psi_n \right)^{N-m} \right], \tag{5.224}$$

where the parameters a_m are *N arbitrary* (possibly *complex*) numbers, the *N* (possibly *complex*) functions $\psi_n(x,t)$ are the dependent variables, and the *N* (possibly *complex*) functions $\varphi_m(x,t)$ and their partial derivatives $\varphi_{m,x}(x,t)$ are expressed in terms of the dependent variables $\psi_n(x,t)$ and their partial derivatives $\psi_{m,x}(x,t)$ by the formulas (5.223). ■

This means that the initial-value problem—to compute the N functions $\psi_n(x, t)$ for all time $t > 0$ from given initial data $\psi_n(x, 0)$—can be solved by *algebraic* operations (including changes of variables from the *coefficients* to the *zeros* of a polynomial of degree N such as (5.222)) and *quadratures*. The procedure to do so is detailed in the following Section 5.3, and this implies the validity of the solutions reported immediately below.

For $N = 2$ this system, (5.224), of 2 coupled nonlinear PDEs reads as follows:

$$\psi_{n,t} + \psi_{n,xx} = (-1)^{n+1} (\psi_1 - \psi_2)^{-1} \big[2\, \psi_{1,x}\, \psi_{2,x}$$
$$+ a_1 \left(\psi_{1,x} + \psi_{2,x} \right) \left(\psi_1 + \psi_2 \right) \psi_n$$
$$- a_2 \left(\psi_{1,x}\, \psi_2 + \psi_1\, \psi_{2,x} \right) \psi_1\, \psi_2 \big] , \quad n = 1, 2 . \quad (5.225)$$

An example of specific solution of this system of 2 coupled nonlinear PDEs, (5.225), reads as follows:

$$\psi_n(x, t) = -\frac{1 + (-1)^n \left\{ 1 - 4 \left[f_1(x, t) \right]^2 / f_2(x, t) \right\}^{1/2}}{2 f_1(x, t)} , \quad n = 1, 2 , \quad (5.226a)$$

$$f_n(x, t) = -\frac{a_n}{2\, \gamma_n} + \beta_n \exp\left[-\gamma_n \left(x - \gamma_n\, t \right) \right] , \quad n = 1, 2 , \quad (5.226b)$$

where the 2 parameters a_n are those appearing in the PDEs (5.225) and the 4 (nonvanishing) parameters β_n and γ_n can be *arbitrarily* assigned. Note that if $\gamma_1 = \gamma_2 = \gamma$ this solution has the "single-soliton" feature to depend on the space and time coordinates only via their combination $x - \gamma t$.

5.2 A S-Integrable System of Nonlinear PDEs

Proposition 5.2.1. The following system of N coupled nonlinear PDEs in $1 + 1$ variables is *S-integrable*:

$$\psi_{n,t} + \psi_{n,xxx} = 3 \sum_{\ell=1, \ell \neq n}^{N} \left(\frac{\psi_{n,xx}\, \psi_{\ell,x} + \psi_{n,x}\, \psi_{\ell,xx}}{\psi_n - \psi_\ell} \right)$$
$$- 3 \sum_{\ell_1, \ell_2 = 1;\, \ell_1 \neq \ell_2,\, \ell_1, \ell_2 \neq n}^{N} \frac{\psi_{n,x}\, \psi_{\ell_1,x}\, \psi_{\ell_2,x}}{\left(\psi_n - \psi_{\ell_1} \right) \left(\psi_n - \psi_{\ell_2} \right)}$$
$$+ \left[\prod_{\ell=1,\, \ell \neq n}^{N} \left(\psi_n - \psi_\ell \right) \right]^{-1} \sum_{m=1}^{N} \left[a_m\, \varphi_{m,x}\, \varphi_m\, \left(\psi_n \right)^{N-m} \right] , \quad (5.227)$$

where the parameters a_m are N arbitrary (*complex*) numbers, the N (*complex*) functions $\psi_n(x, t)$ are the dependent variables, and the N (*complex*) functions $\varphi_m(x, t)$ as well as their partial derivatives $\varphi_{m,x}(x, t)$ are expressed in terms of the dependent variables $\psi_n(x, t)$ and their partial derivatives $\psi_{n,x}(x, t)$ by the formulas (5.223). ■

This means that the initial-value problem—to compute the N functions $\psi_n(x, t)$ for all time $t > 0$ from given initial data $\psi_n(x, 0)$—can be solved by *algebraic* operations (including changes of variables from the *coefficients* to the *zeros* of a polynomial of degree N such as (5.222)) and via the standard *spectral transform* technique (see for instance [46]). The procedure to do so is detailed in the following Section 5.3, and this implies the validity of the solutions reported immediately below.

For $N = 2$ this system of 2 coupled nonlinear PDEs reads as follows:

$$
\begin{aligned}
\psi_{n,t} + \psi_{n,xxx} = (-1)^{n+1} \, (\psi_1 - \psi_2)^{-1} \, & \big\{ 3 \left(\psi_{1,xx} \, \psi_{2,x} + \psi_{1,x} \, \psi_{2,xx} \right) \\
& - \left[a_1 \left(\psi_{1,x} + \psi_{2,x} \right) \left(\psi_1 + \psi_2 \right) \psi_n \right] \\
& + a_2 \left(\psi_{1,x} \, \psi_2 + \psi_1 \, \psi_{2,x} \right) \psi_1 \, \psi_2 \big\} \,, \quad n = 1, 2 \,.
\end{aligned} \tag{5.228}
$$

An example of specific solution of this system of 2 coupled nonlinear PDEs, (5.228), reads as follows:

$$
\begin{aligned}
\psi_n(x, t) = {} & \left\{ \frac{-6 \, \beta_1 \, (\gamma_1)^2}{a_1 \, \cosh^2 \left[\gamma_1 \left(x - 4 \, \gamma_1 \, t \right) \right]} \right\} \cdot \\
& \cdot \left\{ 1 + (-1)^n \left[1 - \frac{(a_1)^2 \, \beta_2 \, (\gamma_2)^2 \, \cosh^4 \left[\gamma_1 \left(x - 4 \, \gamma_1 \, t \right) \right]}{3 \, a_2 \, (\beta_1)^2 \, (\gamma_1)^4 \, \cosh^2 \left[\gamma_2 \left(x - 4 \, \gamma_2 \, t \right) \right]} \right]^{1/2} \right\}, \\
& n = 1, 2 \,,
\end{aligned} \tag{5.229}
$$

where the 2 parameters a_n are those appearing in the system (5.228) and the 4 (nonvanishing) parameters β_n and γ_n can be *arbitrarily* assigned. Note that if these 4 parameters are *all real* it is then *sufficient* that the parameters ratio β_2/a_2 be *negative*, $\beta_2/a_2 < 0$, for this solution to be *real* and *nonsingular* for *all real* values of the dependent variables x and t. Also note that if $\gamma_1 = \gamma_2 = \gamma$, this solution has the "single-soliton" feature to depend on the space and time coordinates only via their combination $x - 4\gamma t$.

For $N = 3$ this system of 3 coupled nonlinear PDEs, (5.227), reads as follows:

$$
\begin{aligned}
\psi_{n,t} + \psi_{n,xxx} = {} & 3 \sum_{s=1,2} \left(\frac{\psi_{n,xx} \, \psi_{n+s,x} + \psi_{n,x} \, \psi_{n+s,xx}}{\psi_n - \psi_{n+s}} \right) \\
& + \left[(\psi_n - \psi_{n+1}) \, (\psi_n - \psi_{n+2}) \right]^{-1} \left\{ -6 \left[\psi_{n,x} \, \psi_{n+1,x} \, \psi_{n+2,x} \right] \right. \\
& \left. + \sum_{m=1}^{3} \left[a_m \, \varphi_{m,x} \, \varphi_m \, (\psi_n)^{N-m} \right] \right\} \,, \quad n = 1, 2, 3 \quad \mathrm{mod}\,[3] \,, \tag{5.230}
\end{aligned}
$$

with $\varphi_m(x, t)$ and $\varphi_{m,x}(x, t)$ given by (5.223) (with $N = 3$).

5.3 Proofs

The proofs of the two Propositions 5.1.1 and 5.2.1 are quite easy. The starting point are the 3 *identities*

$$\psi_{n,t} = -\left[\prod_{\ell=1,\ \ell\neq n}^{N} (\psi_n - \psi_\ell)\right]^{-1} \sum_{m=1}^{N} \left[\varphi_{m,t}\ (\psi_n)^{N-m}\right], \qquad (5.231a)$$

$$\psi_{n,xx} = \sum_{\ell=1,\ \ell\neq n}^{N} \left(\frac{2\ \psi_{n,x}\ \psi_{\ell,x}}{\psi_n - \psi_\ell}\right)$$

$$-\left[\prod_{\ell=1,\ \ell\neq n}^{N} (\psi_n - \psi_\ell)\right]^{-1} \sum_{m=1}^{N} \left[\varphi_{m,xx}\ (\psi_n)^{N-m}\right], \qquad (5.231b)$$

$$\psi_{n,xxx} = 3 \sum_{\ell=1,\ell\neq n}^{N} \left(\frac{\psi_{n,xx}\ \psi_{\ell,x} + \psi_{n,x}\ \psi_{\ell,xx}}{\psi_n - \psi_\ell}\right)$$

$$-3 \sum_{\ell_1,\ell_2=1,\ \ell_1\neq\ell_2,\ \ell_1,\ell_2\neq n}^{N} \left[\frac{\psi_{n,x}\ \psi_{\ell_1,x}\ \psi_{\ell_2,x}}{(\psi_n - \psi_{\ell_1})\ (\psi_n - \psi_{\ell_2})}\right]$$

$$-\left[\prod_{\ell=1,\ \ell\neq n}^{N} (\psi_n - \psi_\ell)^{-1}\right] \sum_{m=1}^{N} \left[\varphi_{m,xxx}\ (\psi_n)^{N-m}\right], \qquad (5.231c)$$

which relate the *N zeros* ψ_n and the *N coefficient* φ_m of a polynomial such as (5.222) (and correspond—up to notational changes—to the 3 *identities* (2.7), (2.8) and (2.9)).

We now note that the sum of the first two of these three *identities* (5.231) imply the *identity*

$$\psi_{n,t} + \psi_{n,xx} = \sum_{\ell=1,\ \ell\neq n}^{N} \left(\frac{2\ \psi_{n,x}\ \psi_{\ell,x}}{\psi_n - \psi_\ell}\right)$$

$$-\left[\prod_{\ell=1,\ \ell\neq n}^{N} (\psi_n - \psi_\ell)\right]^{-1} \sum_{m=1}^{N} \left\{\left[\varphi_{m,t} + \varphi_{m,xx}\right]\ (\psi_n)^{N-m}\right\}; \qquad (5.232a)$$

and likewise the sum of the first and third of the *identities* (5.231) implies the *identity*

$$\psi_{n,t} + \psi_{n,xxx} = 3 \sum_{\ell=1, \ell \neq n}^{N} \left(\frac{\psi_{n,xx} \, \psi_{\ell,x} + \psi_{n,x} \, \psi_{\ell,xx}}{\psi_n - \psi_\ell} \right)$$

$$- 3 \sum_{\ell_1, \ell_2 = 1, \, \ell_1 \neq \ell_2, \, \ell_1, \ell_2 \neq n}^{N} \left[\frac{\psi_{n,x} \, \psi_{\ell_1,x} \, \psi_{\ell_2,x}}{(\psi_n - \psi_{\ell_1}) \, (\psi_n - \psi_{\ell_2})} \right]$$

$$- \left[\prod_{\ell=1, \, \ell \neq n}^{N} (\psi_n - \psi_\ell)^{-1} \right] \sum_{m=1}^{N} \left\{ [\varphi_{m,t} + \varphi_{m,xxx}] \, (\psi_n)^{N-m} \right\} .$$

$$(5.232b)$$

Now assume that the *N functions* $\varphi_m(x,t)$ satisfy the Burgers equations

$$\varphi_{m,t} + \varphi_{m,xx} = a_m \, \varphi_{m,x} \, \varphi_m ; \qquad (5.233)$$

this implies, see (5.232a), that the N functions $\psi_n(x,t)$ satisfy the system of PDEs (5.224). Proposition 5.1.1 is thereby proven. And the solution of the initial-value problem for this system of PDEs, (5.224), is yielded by the following procedure. *Step (i)*: from the initial data $\psi_n(x,0)$, compute the corresponding functions $\varphi_m(x,0)$ (via the formulas (5.223)). *Step (ii)*: solve the *C-integrable* PDEs (5.233) with these initial data $\varphi_m(x,0)$, obtaining thereby the functions $\varphi_m(x,t)$ for all time $t > 0$. *Step (iii)*: the solutions $\psi_n(x,t)$ of the system of PDEs (5.224) are then provided by the N zeros of the polynomial (5.222) with *coefficients* $\varphi_m(x,t)$. The explicit solution (5.226) is manufactured using the well known single-soliton solutions of the Burgers equations (5.233).

The proof of Proposition 5.2.1, and the procedure to solve the system of PDEs (5.227), are quite analogous, except that the role of the *identity* (5.232a) is now played by the *identity* (5.232b), and the role played by the *C-integrable* Burgers PDEs (5.233) is now played by the *S-integrable* KdV PDEs

$$\varphi_{m,t} + \varphi_{m,xxx} + a_m \, \varphi_{m,x} \, \varphi_m = 0 . \qquad (5.234)$$

Likewise for the manufacture of the explicit solution (5.229).

5.4 Outlook

The approach described above (in Chapter 5) provides the possibility to identify a large universe of *new integrable/solvable systems of nonlinear PDEs*; the two PDEs specifically discussed in Sections 5.1 and 5.2 are merely examples of the vistas opened by this methodology to identify *new integrable/solvable* systems of nonlinear PDEs. Note, for instance, that the assumptions made above—that all the *coefficients* $\varphi_m(x,t)$ satisfy the *same* integrable/solvable PDE (up to a change of its parameters)—are *not* necessary; for instance in the case of Proposition 5.1.1, some of the *N* coefficients $\varphi_m(x,t)$ might satisfy, say, the *C-integrable* Kundu-Eckhaus PDE [80], [48] instead of the Burgers PDE, and in the case of Proposition 5.2.1 some of the *N* coefficients $\varphi_m(x,t)$ might satisfy, say, the *S-integrable* Modified KdV PDE (see, for instance, [46]) instead of the KdV PDE; and it is also possible to let some of the *N* coefficients $\varphi_m(x,t)$ evolve according to *C-integrable* PDEs and others evolve according to *S-integrable* PDEs ... Moreover, all the novel *integrable/solvable* PDEs identified via this approach can themselves be subsequently interpreted as characterizing the evolution of the *N* *coefficients* of a monic polynomial of degree *N*, hence as inputs for the *generation* (see the next Chapter 6), via this approach, of *hierarchies* of new systems of *integrable/solvable* PDEs; and it is also possible to extend this approach to a *multidimensional* context (i.e., beyond the $(1 + 1)$-dimensional context to which the treatment in Chapter 5 has been restricted) [18], [69] and to more general auxiliary functions than polynomials [51] [52].

Let us end this Section 5.4 by displaying a system of *N* *nonlinearly* coupled PDEs the solution of which can be *explicitly* exhibited for arbitrary initial data (up to *algebraic* operations, *explicitly* performable for $N \leq 4$); the proof of these results being left as an exercise for the interested reader (hint: $\varphi_{m,tt} = v_m^2 \varphi_{m,xx}$). This system of PDEs reads as follows:

$$
\psi_{n,tt} = \sum_{\ell=1,\,\ell\neq n}^{N} \left(\frac{2\,\psi_{n,t}\,\psi_{\ell,t}}{\psi_n - \psi_\ell} \right) - \left[\prod_{\ell=1,\,\ell\neq n}^{N} (\psi_n - \psi_\ell) \right]^{-1} \cdot
$$

$$
\cdot \sum_{m=1}^{N} \left[(-1)^m\,(v_m)^2 \sum_{1\leq n_1 < n_2 < \cdots < n_m \leq N} \left(\psi_{n_1} \cdot \psi_{n_2} \cdots \psi_{n_m} \right)_{xx} (\psi_n)^{N-m} \right],
$$
$$(5.235a)$$

and its solution is provided by the *N zeros* $\psi_n(x,t)$ of the polynomial (5.222) with its *N coefficients* $\varphi_m(x,t)$ given by the following *explicit* formulas

$$
\varphi_m(x,\,t) = (2\,v_m)^{-1} \left\{ v_m \left[\varphi_m(x + v_m\,t,\,0) + \varphi_m(x - v_m\,t,\,0) \right] \right.
$$
$$
\left. + \varphi_{m,t}(x + v_m\,t,\,0) - \varphi_{m,t}(x - v_m\,t,\,0) \right\} \qquad (5.235b)
$$

in terms of the *initial* data $\varphi_m(x,\,0)$ and $\varphi_{m,t}(x,\,0)$, themselves expressed by the *explicit* formulas (5.223) in terms of the *initial* data $\psi_m(x,\,0)$ and $\psi_{m,t}(x,\,0)$ associated to the system (5.235a). Here the *N* parameters v_m are *N arbitrary nonvanishing* numbers.

Remark 5.4.1.   If the N parameters v_m are *all rational multiples* r_m of the same *nonvanishing real* number v,

$$v_m = r_m \, v \,, \tag{5.236a}$$

then the system of *nonlinear* PDEs (5.235a) features the following remarkable property: if the *initial* data $\psi_n(x,\, 0)$ respectively $\psi_{n,t}(x,\, 0)$ are *periodic* functions of their arguments x with periods which are *all rational multiples* $r_m^{(1)}$ respectively $r_m^{(2)}$ of a basic spacing $\bar{x}$,

$$\psi_m\left(x + \bar{x}_m^{(1)},\, 0\right) = \psi_m(x,\, 0) \,, \quad \psi_{m,t}\left(x + \bar{x}_m^{(2)},\, 0\right) = \psi_{m,t}(x,\, 0) \,,$$
$$\bar{x}_m^{(1)} = r_m^{(1)} \, \bar{x} \,, \quad \bar{x}_m^{(2)} = r_m^{(2)} \, \bar{x} \,, \tag{5.236b}$$

then for *all* time the corresponding solutions $\psi_n(x,\, t)$ are as well *periodic* in *both* space and time,

$$\psi_n\left(x + v_1 \, \bar{x},\, t + v_2 \, \frac{\bar{x}}{|v|}\right) = \psi_n(x,\, t) \,, \tag{5.236c}$$

with v_1 and v_2 appropriate *positive integers*. ■

For $N = 2$ the system (5.235a) reads as follows:

$$\psi_{n,tt} = \frac{(-1)^{n+1}}{\psi_1 - \psi_2} \Big[2\, \psi_{1,t}\, \psi_{2,t} + (v_1)^2 \, \left(\psi_{1,xx} + \psi_{2,xx}\right)\, \psi_n$$
$$- (v_2)^2 \, \left(\psi_{1,xx}\, \psi_2 + \psi_{2,xx}\, \psi_1 + 2\, \psi_{1,x}\, \psi_{2,x}\right) \Big] \,, \, n = 1, 2 \,. \tag{5.237}$$

5.N Notes on Chapter 5

For the findings reported in Chapter 5, see [43]. The idea to identify systems of nonlinear PDEs *amenable to exact treatment* by looking at the evolution of the *zeros* of polynomials the *coefficients* of which evolve as PDEs *amenable to exact treatments* was already exploited in the past, see for instance the 4 papers [18], [69], [51], [52]; but it was limited there to cases with the *coefficients* evolving *linearly*.

6

Generations of Monic Polynomials

In Chapter 6 we describe the notion of *generations of monic polynomials,* such that the *coefficients* of the polynomials of the *next generation* coincide with the *zeros* of the polynomials of the *current generation*. We then indicate how this concept allows to identify *endless hierarchies* of N-body problems *solvable* by *algebraic* operations. The solution of each N-body problem in such an endless sequence coincides with the time evolution in the *complex* plane of the N *zeros* of a time-dependent monic polynomial of degree N belonging to an appropriate *generation* of such polynomials.

While an important motivation for the introduction of this notion of *generations of polynomials* is because it is instrumental for the just mentioned identification of *endless hierarchies* of *new N*-body problems, this notion deserves further study also due to its *natural/unnatural* character in the context of the study of polynomials. Indeed, a monic polynomial is characterized by its N *coefficients* as well as by its N *zeros*, so a sequence of polynomials in which these two sets of numbers exchange sequentially their roles is an *interesting* possibility; note that interchanging the *coefficients* and the *zeros* of polynomials seems *unorthodox* given the different nature of these two sets of numbers—one *ordered* and the other one *unordered*— but this fact makes this possibility more *intriguing* (indeed, an interesting problem in this connection was raised over half a century ago by S. M. Ulam; see below Section 6.2).

The main protagonists of this Chapter 6—and, indeed, of this book—are the time-dependent monic polynomials of degree N in the variable z, which we write again (see (2.3)) in terms of their *coefficients* y_m and their *zeros* x_n:

$$p_N(z; \vec{y}; \tilde{x}) = z^N + \sum_{m=1}^{N} \left[y_m \, z^{N-m} \right] , \qquad (6.238a)$$

$$p_N(z; \vec{y}; \tilde{x}) = \prod_{n=1}^{N} [z - x_n] . \qquad (6.238b)$$

And let us again emphasize that this notation is somewhat redundant, since this polynomial (6.238) is equally well identified by the N-vector $\vec{y}$ the N components of which are its N *coefficients* y_m, as by the *unordered* set $\tilde{x}$ the N elements of which are its N zeros x_n. Indeed the N coefficients y_m can be *explicitly* expressed in terms of the N zeros x_n, see (2.4); while the N zeros x_n are likewise *uniquely* determined—but only *up to permutations*—by the N coefficients y_m, although *explicit* expressions to this effect are generally available only for $N \leq 4$.

>   **Remark 6.1.**   Above and hereafter we assume to be in the *generic* case when the
> N zeros x_n of the Nth degree polynomials under consideration are *all different among*
>   *themselves*, $x_n \neq x_\ell$ if $n \neq \ell$. ∎

To introduce the notion of *generations* of monic polynomials such that the *coefficients* of the polynomials of the *next generation* coincide with the *zeros* of the polynomials of the *current generation*, let us start from a *seed* (monic) polynomial of degree N and use its N zeros to construct the polynomials of the *first generation* associated with this *seed*: the N *coefficients* of these *first-generation* polynomials are the N zeros of the *seed* polynomial. Since there are $N!$ ways to order the N zeros of the *seed* polynomial, there are $N!$ *different* (see Remark 6.1) polynomials in the *first* generation. We then repeat this process for each of the $N!$ different monic polynomials in the *first generation*, to construct the *second generation* of monic polynomials. Because each polynomial from the *first* generation yields $N!$ polynomials in the *second generation* (in the *generic* case when the *zeros* of all the polynomials under consideration are distinct, see Remark 6.1), there are $(N!)^2$ polynomials in the *second generation*. And by iterating this process an *endless hierarchy* of monic polynomials of degree N is created. There are $(N!)^k$ polynomials in the k-th *generation* of this *hierarchy*, which can be visualized as an *infinite rooted tree*. The *seed* polynomial of the hierarchy is the *root* of this *tree*; each *vertex* of the *tree* corresponds to a monic polynomial and has $N!$ upper branches. More precisely, if a vertex v corresponds to a monic polynomial of degree N, then its $N!$ children vertices correspond to the $N!$ monic polynomials of degree N such that their N *coefficients* coincide with one of the $N!$ permutations of the N *zeros* of this polynomial.

While the idea to manufacture such infinite polynomial *hierarchies* is simple, the description of the $(N!)^k$ polynomials in each generation of the hierarchy requires some notational care. Indeed, in order to identify uniquely every polynomial in every *generation* of the hierarchy, we must keep track of both the *coefficients* and the *zeros* of every polynomial of the hierarchy. We already mentioned that such notation is somewhat redundant (see the remark after (6.238)); but this redundancy is now essential to define precisely the *hierarchy* of polynomials.

To denote a polynomial $p_N(z)$ in the k-th generation of a hierarchy constructed using the above process in a way that distinguishes this polynomial from the other polynomials in the hierarchy we must therefore introduce a multi-index $\vec{\mu}^{(k)}$ in formulas analogous to (6.238). This multi-index $\vec{\mu}^{(k)}$ is a k-vector with k integer components $\mu_1, \mu_2, \ldots, \mu_k$, which identify the *path* in the *tree* that characterizes the *hierarchy*, connecting the polynomial $p_N(z)$ indexed by $\vec{\mu}^{(k)}$ with the *root* of the *tree*—the *seed* polynomial. This path, and therefore the multi-index $\vec{\mu}^{(k)}$, indicates the choices of the orderings of the *zeros* of each polynomial in each of the generations prior to the k-th generation—choices that lead to the polynomial $p_N(z)$ in the k-th generation identified by its multi-index $\vec{\mu}^{(k)}$. Therefore we generalize notation (6.238) by defining the polynomials

$$p_N^{(\vec{\mu}^{(k)})}\left(z;\ \vec{y}^{(\vec{\mu}^{(k)})};\ \tilde{x}^{(\vec{\mu}^{(k)})}\right) = z^N + \sum_{m=1}^{N}\left[y_m^{(\vec{\mu}^{(k)})}\ z^{N-m}\right],\qquad(6.239a)$$

$$p_N^{(\vec{\mu}^{(k)})}\left(z;\ \vec{y}^{(\vec{\mu}^{(k)})};\ \tilde{x}^{(\vec{\mu}^{(k)})}\right) = \prod_{n=1}^{N}\left[z - x_n^{(\vec{\mu}^{(k)})}(t)\right].\qquad(6.239b)$$

Here and below k is a *nonnegative integer* taking the values $0, 1, 2, \ldots$, which characterizes the generations of polynomials, while $\vec{\mu}^{(k)}$ is a k-vector the k components $\mu_1^{(k)}, \mu_2^{(k)}, \ldots, \mu_k^{(k)}$ of which are *integers* in the range from 1 to $N!$, the significance of which was explained above and is detailed below.

Let us start from the *seed polynomial* characterized by the index $k = 0$, for which we use the following notation:

$$p_N^{(0)}\left(z;\ \vec{y}^{(0)};\ \tilde{x}^{(0)}\right) = z^N + \sum_{m=1}^{N}\left[y_m^{(0)}\ z^{N-m}\right],\qquad(6.240a)$$

$$p_N^{(0)}\left(z;\ \vec{y}^{(0)};\ \tilde{x}^{(0)}\right) = \prod_{n=1}^{N}\left[z - x_n^{(0)}\right].\qquad(6.240b)$$

The N zeros $x_n^{(0)}$ of this *seed* polynomial—i.e., the N elements of the *unordered* set $\tilde{x}^{(0)}$—can be ordered in $N!$ ways to produce the N-vectors $\vec{x}_{[\mu_1]}^{(0)}$, where the index μ_1, ranging from 1 to $N!$, identifies a particular ordering. For example, $\vec{x}_{[1]}^{(0)}$ can be the N-vector the N components of which are the N elements of the unordered set $\tilde{x}^{(0)}$ arranged in *lexicographic* order (say, of any pair of different *complex* numbers, that with *smaller real part* comes *first*, and *if the real parts are equal* that with *smaller imaginary part* comes *first*), and the remaining N-vectors $\vec{x}_{[2]}^{(0)}, \ldots, \vec{x}_{[N!]}^{(0)}$ are then obtained from $\vec{x}_{[1]}^{(0)}$ by *permuting* the components of this vector according to the *lexicographic* order of *permutations*. Note that hereafter the n-th component of the vector $\vec{x}^{(0)}_{[\mu_1]}$ is denoted by $x^{(0)}_{[\mu_1],n}$.

The $N!$ polynomials of the *first generation* ($k = 1$) are then defined by the following formulas analogous to (6.239):

$$p_N^{(\mu_1)}\left(z; \vec{y}^{(\mu_1)}; \tilde{x}^{(\mu_1)}\right) = z^N + \sum_{m=1}^{N}\left[y_m^{(\mu_1)}\, z^{N-m}\right], \qquad (6.241a)$$

$$p_N^{(\mu_1)}\left(z; \vec{y}^{(\mu_1)}; \tilde{x}^{(\mu_1)}\right) = \prod_{n=1}^{N}\left[z - x_n^{(\mu_1)}\right], \qquad (6.241b)$$

but now with

$$\vec{y}^{(\mu_1)} = \vec{x}^{(0)}_{[\mu_1]}, \quad y_m^{(\mu_1)} = x^{(0)}_{[\mu_1],m} \qquad (6.241c)$$

where $1 \le \mu_1 \le N!$. Note that then to each of the $N!$ polynomials $p_N^{(\mu_1)}(z)$ belonging to the *first generation* is associated a specific *unordered* set of its N zeros $\tilde{x}^{(\mu_1)}$, see (6.241b).

Example 6.1.   As simple example let us take the following polynomial of degree $N = 2$ as *seed*:

$$p_2^{(0)}(z) = z^2 + y_1^{(0)}\, z + y_2^{(0)} = \left(z - x_1^{(0)}\right)\left(z - x_2^{(0)}\right). \qquad (6.242a)$$

Then the $2! = 2$ polynomials of the *first* ($k = 1$) generation are

$$p_2^{(1)}(z) = z^2 + x_1^{(0)}\, z + x_2^{(0)} = \left(z - x_1^{(1)}\right)\left(z - x_2^{(1)}\right), \qquad (6.242b)$$

$$p_2^{(2)}(z) = z^2 + x_2^{(0)}\, z + x_1^{(0)} = \left(z - x_1^{(2)}\right)\left(z - x_2^{(2)}\right). \qquad (6.242c)$$

Since we have restricted this example to the case of a second-degree polynomial ($N = 2$), the two *zeros* $x_1^{(0)}$ and $x_2^{(0)}$ of the *seed* polynomial (6.242a) could be *explicitly* expressed in terms of the two *coefficients* $y_1^{(0)}$ and $y_2^{(0)}$ of this *seed* polynomial. Likewise, the two *zeros* $x_1^{(1)}$ and $x_2^{(1)}$ of the *first generation* polynomial $p_2^{(1)}(z)$ could be *explicitly* expressed in terms of the two *coefficients* $x_1^{(0)}$ and $x_2^{(0)}$ of the polynomial $p_2^{(1)}(z)$ hence also in terms of the two *coefficients* $y_1^{(0)}$ and $y_2^{(0)}$ of the *seed* polynomial (6.242a). And a similar comment can also be made about the two *zeros* $x_1^{(2)}$ and $x_2^{(2)}$ of the other *first generation* polynomial $p_2^{(2)}(z)$. ∎

It is now evident how this procedure can be continued, producing thereby a *tree* of polynomials. But, in view of the novelty of this notion, let us be pedantic and describe again in detail the *next generation*, and then how to go from *any generation* to the *next* one.

The N zeros $x_n^{(\mu_1)}$ of a *first-generation* polynomial $p_N^{(\mu_1)}(z)$, see (6.241), can be ordered in $N!$ ways to produce the N-vectors $\vec{x}_{[\mu_2]}^{(\mu_1)}$, where the index μ_2 ranging from 1 to $N!$ indicates a particular ordering of the N zeros $x_n^{(\mu_1)}$ as components of these N-vectors. The n-th component of the vector $\vec{x}_{[\mu_2]}^{(\mu_1)}$ is then denoted as $x_{[\mu_2],n}^{(\mu_1)}$.

Then the next ($k = 2$) *second generation* of polynomials is defined as follows:

$$p_N^{(\vec{\mu}^{(2)})}\left(z;\ \vec{y}^{(\vec{\mu}^{(2)})};\ \tilde{x}^{(\vec{\mu}^{(2)})}\right) = z^N + \sum_{m=1}^{N}\left[y_m^{(\vec{\mu}^{(2)})}\ z^{N-m}\right]\,, \tag{6.243a}$$

$$p_N^{(\mu^{(2)})}\left(z;\ \vec{y}^{(\vec{\mu}^{(2)})};\ \tilde{x}^{(\vec{\mu}^{(2)})}\right) = \prod_{n=1}^{N}\left[z - x_n^{(\mu^{(2)})}\right]\,, \tag{6.243b}$$

with

$$\vec{y}^{(\vec{\mu}^{(2)})} = \vec{x}_{[\mu_2]}^{(\mu_1)}\,,\quad y_m^{(\vec{\mu}^{(2)})} = x_{[\mu_2],m}^{(\mu_1)}\,, \tag{6.243c}$$

where $\vec{\mu}^{(2)}$ is a multi-index 2-vector, $\vec{\mu}^{(2)} = (\mu_1, \mu_2)$, with its 2 components μ_1, μ_2 taking each $N!$ *integer* values in the range from 1 to $N!$, $1 \leq \mu_1, \mu_2 \leq N!$. Each of the $(N!)^2$ polynomials $p_N^{(\vec{\mu}^{(2)})}$ in the *second generation* is thus characterized by its multi-index $\vec{\mu}^{(2)}$ hence by its N *coefficients* $y_m^{(\vec{\mu}^{(2)})}$ and also by the *unordered* set $\tilde{x}^{(\vec{\mu}^{(2)})}$ of its N zeros.

It is now evident how all subsequent generations of polynomials can be manufactured, all of them being generated by the initial *seed* polynomial (6.240). To make the matter completely clear, let us indicate how the *k-generation* polynomials are obtained from the preceding $(k-1)$-*generation* polynomials.

The N zeros $x_n^{(\vec{\mu}^{(k-1)})}$ of a polynomial $p_N^{(\vec{\mu}^{(k-1)})}(z)$ belonging to the $(k-1)$-*generation* can be ordered in $N!$ ways to produce the N-vectors $\vec{x}_{[\mu_k]}^{(\vec{\mu}^{(k-1)})}$, where the index μ_k—the last *integer* component of the multi-index k-vector $\vec{\mu}^{(k)}$, with values ranging from 1 to $N!$—identifies a particular ordering of the N zeros $x_n^{(\vec{\mu}^{(k-1)})}$. The n-th component of the N-vector $\vec{x}_{[\mu_k]}^{(\vec{\mu}^{(k-1)})}$ is then denoted as $x_{[\mu_k],n}^{(\vec{\mu}^{(k-1)})}$.

The polynomials belonging to the k-th generation are then defined as follows:

$$p_N^{(\vec{\mu}^{(k)})}\left(z;\ \vec{y}^{(\vec{\mu}^{(k)})};\ \tilde{x}^{(\vec{\mu}^{(k)})}\right) = z^N + \sum_{m=1}^{N}\left[y_m^{(\vec{\mu}^{(k)})}\ z^{N-m}\right]\,, \tag{6.244a}$$

$$p_N^{(\vec{\mu}^{(k)})}\left(z;\ \vec{y}^{(\vec{\mu}^{(k)})};\ \tilde{x}^{(\vec{\mu}^{(k)})}\right) = \prod_{n=1}^{N}\left[z - x_n^{(\vec{\mu}^{(k)})}\right]\,, \tag{6.244b}$$

with

$$\vec{y}^{(\vec{\mu}^{(k)})} = \vec{x}_{[\mu_k]}^{(\vec{\mu}^{(k-1)})}\,,\quad y_m^{(\vec{\mu}^{(k)})} = x_{[\mu_k],m}^{(\vec{\mu}^{(k-1)})}\,, \tag{6.244c}$$

where $\vec{\mu}^{(k)} = (\vec{\mu}^{(k-1)}, \mu^k) = (\mu_1, \ldots, \mu_{k-1}, \mu_k)$ is a multi-index k-vector with each of its k components $\mu_1, \ldots, \mu_k$ being an *integer* in the range from 1 to $N!$. Note that the polynomials shown in (6.244) identifies $(N!)^k$ different polynomials in the k-th *generation*, each polynomial $p_N^{(\vec{\mu}^{(k)})}(z)$ then identifying a specific *unordered* set of N *zeros* $\tilde{x}^{\vec{\mu}^{(k)}}$ see (6.244b).

Let us end this first part of Chapter 6 by displaying the first 3 *generations* of second-degree (monic) polynomials, constructed using the procedure described above starting from the generic *seed* polynomial

$$p_2^{(0)}(z) = z^2 + bz + c, \tag{6.245}$$

where b and c are two *arbitrary complex* numbers.

Before proceeding, let us note that every *complex* number $\zeta = \rho \exp(\mathbf{i}\,\phi)$—where ρ is a *nonnegative real* number and ϕ is a *real* number in the range $0 \leq \phi < 2\pi$—has two square roots $\pm r$, where $r = \sqrt{\rho} \exp(\mathbf{i}\,\phi/2)$ and, for definiteness, the square root $\sqrt{\rho}$ is *nonnegative*. Using this fact, we introduce the (generally *complex*) numbers $r_0, r_{11}, r_{12}, r_{21}, r_{22}, r_{23}, r_{24}$ as follows:

$$\sqrt{b^2 - 4c} = \pm r_0,$$
$$\sqrt{8b + 2b^2 - 4c + 8r_0 - 2br_0} = \pm r_{11},$$
$$\sqrt{8b + 2b^2 - 4c - 8r_0 + 2br_0} = \pm r_{12},$$
$$\sqrt{-8b + 4b^2 - 8c + 24r_0 - 4br_0 + 16r_{11} + 2br_{11} - 2r_0r_{11}} = \pm r_{21},$$
$$\sqrt{-8b + 4b^2 - 8c + 24r_0 - 4br_0 - 16r_{11} - 2br_{11} + 2r_0r_{11}} = \pm r_{22},$$
$$\sqrt{-8b + 4b^2 - 8c - 24r_0 + 4br_0 + 16r_{12} + 2br_{12} + 2r_0r_{12}} = \pm r_{23},$$
$$\sqrt{-8b + 4b^2 - 8c - 24r_0 + 4br_0 - 16r_{12} - 2br_{12} - 2r_0r_{12}} = \pm r_{24}. \tag{6.246}$$

Using this notation, we display below the polynomials in the first few generations, yielded by the *seed* polynomial (6.245).

First generation. The 2 polynomials of the *first generation* read:

$$p_2^{(1)}(z) = z^2 - \frac{b}{2}(z+1) + \frac{r_0}{2}(z-1),$$
$$p_2^{(2)}(z) = z^2 - \frac{b}{2}(z+1) - \frac{r_0}{2}(z-1). \tag{6.247a}$$

Second generation. The 4 polynomials of the *second generation* read:

$$p_2^{(1,1)}(z) = z^2 + \frac{b - r_0}{4}(z+1) + \frac{r_{11}}{4}(z-1),$$
$$p_2^{(1,2)}(z) = z^2 + \frac{b - r_0}{4}(z+1) - \frac{r_{11}}{4}(z-1),$$

$$p_2^{(2,1)}(z) = z^2 + \frac{b + r_0}{4}(z + 1) + \frac{r_{12}}{4}(z - 1),$$

$$p_2^{(2,2)}(z) = z^2 + \frac{b + r_0}{4}(z + 1) - \frac{r_{12}}{4}(z - 1). \tag{6.247b}$$

Third generation. The 8 polynomials of the *third generation* read:

$$p_2^{(1,1,1)}(z) = z^2 + \frac{1}{8}(-b + r_0 - r_{11})(z + 1) + \frac{r_{21}}{8}(z - 1),$$

$$p_2^{(1,1,2)}(z) = z^2 + \frac{1}{8}(-b + r_0 - r_{11})(z + 1) - \frac{r_{21}}{8}(z - 1),$$

$$p_2^{(1,2,1)}(z) = z^2 + \frac{1}{8}(-b + r_0 + r_{11})(z + 1) + \frac{r_{22}}{8}(z - 1),$$

$$p_2^{(1,2,2)}(z) = z^2 + \frac{1}{8}(-b + r_0 + r_{11})(z + 1) - \frac{r_{22}}{8}(z - 1),$$

$$p_2^{(2,1,1)}(z) = z^2 + \frac{1}{8}(-b - r_0 - r_{12})(z + 1) + \frac{r_{23}}{8}(z - 1),$$

$$p_2^{(2,1,2)}(z) = z^2 + \frac{1}{8}(-b - r_0 - r_{12})(z + 1) - \frac{r_{23}}{8}(z - 1),$$

$$p_2^{(2,2,1)}(z) = z^2 + \frac{1}{8}(-b - r_0 + r_{12})(z + 1) + \frac{r_{24}}{8}(z - 1),$$

$$p_2^{(2,2,2)}(z) = z^2 + \frac{1}{8}(-b - r_0 + r_{12})(z + 1) - \frac{r_{24}}{8}(z - 1). \tag{6.247c}$$

6.1 Hierarchies of Solvable *N*-Body Problems Related to Generations of Polynomials

In Section 6.1 we indicate how to identify *endless* sequences of *solvable N*-body problems that involve the *generations* of monic polynomials discussed in the preceding part of this Chapter 6. These *N*-body problems are characterized by equations of motion of Newtonian type ("accelerations equals forces"), describing the motion in the *complex* plane of *N* unit-mass point-particles interacting among themselves with highly *nonlinear* forces depending on their positions and their velocities. These *N*-body models—as most of those treated in this book—are "of goldfish type", namely characterized by Newtonian equations of motion reading as follows:

$$\ddot{x}_n = \sum_{\ell=1, \, \ell \neq n}^{N} \left(\frac{2 \, \dot{x}_n \, \dot{x}_\ell}{x_n - x_\ell} \right) + F_n\left(\vec{x}, \dot{\vec{x}}\right), \tag{6.248}$$

where the N (generally *complex*) coordinate $x_n(t)$ identifies the position in the *complex* x-plane at time t of the n-th unit-mass particle, superimposed dots denote time-derivatives, and $F_n(\vec{x}, \dot{\vec{x}})$ are appropriately assigned functions of the N-vectors $\vec{x}(t)$ respectively $\dot{\vec{x}}(t)$ with components $x_n(t)$ respectively $\dot{x}_n(t)$. The "goldfish type" connotation is characterized—as already mentioned above, see Section 4.1—by the presence of the first term in the right-hand side of these equations of motion (6.248), i.e., the sum $\sum_{\ell=1,\ \ell \neq n}^{N} \left[2\dot{x}_n \dot{x}_\ell / (x_n - x_\ell) \right]$.

Let us also recall (see Remark 2.1.1) that an N-body model is considered *solvable* if the configuration of the system at any arbitrary time t can be obtained—for given *initial* data: the *initial* positions and velocities of the N particles in the *complex* x-plane—by *algebraic* operations, such as finding the N zeros $x_n(t)$ of an *explicitly* known time-dependent polynomial of degree N; and let us also re-emphasize that knowledge of the *configuration* of the N-body system at time t—i.e., of the (generally *complex*) values of the N coordinates $x_n(t)$ given as the *unordered* set of the N zeros of a known polynomial of degree N—does *not* allow to identify each specific coordinate, say, the coordinate $x_1(t)$ that has evolved continuously over time from the initial data $x_1(0)$, $\dot{x}_1(0)$: this additional information can only be gained by following over time the evolution of the system, either by integrating numerically the equations of motion, or by identifying the configurations of the system (as given by the N zeros of a polynomial) at a sequence of time intervals sufficiently close to each other so as to guarantee the identification by *contiguity* of the trajectory of each coordinate (or at least of the specific coordinate under consideration); but these additional operations need not be performed with great accuracy, even when one wishes the final configuration—including the identity of each particle—to be known with much greater accuracy.

Likewise—in the case of systems that have been identified as *isochronous* because their solution is provided by the N zeros $x_n(t)$ of a time-dependent polynomial of degree N in z which is itself *periodic* in time with period, say, T—an analogous procedure must be followed to ascertain whether (due to a periodic exchange of the correspondence between the *zeros* of the polynomial and their identities associated to their time evolution) the period of the time evolution of a specific *zero* is T, or νT (with ν a *positive integer* certainly not larger than $\nu_{\text{Max}}(N)$, indeed generally smaller: see Section 2.3).

The key formulas for the following developments are the *identities* (see in particular Subsection 2.2.1) relating the time evolution of the *coefficients* $x_n(t)$ of the *zeros* of a time-dependent (monic) polynomial to that of the *coefficients* $y_m(t)$ of the same polynomial, combined with the notion of *generations* of polynomials discussed in the previous part of this Chapter 6.

The starting point of our treatment is any one of the many known *solvable* N-body models—see for instance [20], [27] and the results reported above in Chapter 4 and below for specific examples—characterized by (Newtonian) equations of motion which we write as follows:

$$\ddot{x}_n = f_n\left(\vec{x}, \dot{\vec{x}}\right) . \tag{6.249}$$

We then consider the generations of (time-dependent) polynomials of degree N in z originating from the (time-dependent) *seed* polynomial

$$p_N^{(0)}\left(z; \vec{y}^{(0)}(t); \tilde{x}^{(0)}(t)\right) = z^N + \sum_{m=1}^{N}\left[y_m^{(0)}(t)\, z^{N-m}\right] , \tag{6.250a}$$

$$p_N^{(0)}\left(z; \vec{y}^{(0)}(t); \tilde{x}^{(0)}(t)\right) = \prod_{n=1}^{N}\left[z - x_n^{(0)}(t)\right] , \tag{6.250b}$$

the (time-dependent) *zeros* $x_n^{(0)}(t)$ of which are the solution of the *solvable* model (6.249):

$$x_n^{(0)}(t) = x_n(t) . \tag{6.250c}$$

It is then evident via (2.8) that the zeros $x_n^{(\mu_1)}(t)$ of the *first generation* polynomials (see (6.241c) and (6.250c))

$$p_N^{(\mu_1)}\left(z; \vec{x}_{[\mu_1]}^{(0)}(t); \tilde{x}^{(\mu_1)}(t)\right) = z^N + \sum_{m=1}^{N}\left[x_{[\mu_1],m}^{(0)}(t)\, z^{N-m}\right] , \tag{6.251a}$$

$$p_N^{(\mu_1)}\left(z; \vec{x}_{[\mu_1]}^{(0)}(t); \tilde{x}^{(\mu_1)}(t)\right) = \prod_{n=1}^{N}\left[z - x_n^{(\mu_1)}(t)\right] , \tag{6.251b}$$

provide the solutions of the N-body problems

$$\ddot{x}_n^{(\mu_1)} = \sum_{\ell=1,\,\ell\neq n}^{N}\left(\frac{2\,\dot{x}_n^{(\mu_1)}\,\dot{x}_\ell^{(\mu_1)}}{x_n^{(\mu_1)} - x_\ell^{(\mu_1)}}\right) - \left[\prod_{\ell=1,\,\ell\neq n}^{N}\left(x_n^{(\mu_1)} - x_\ell^{(\mu_1)}\right)^{-1}\right] \cdot$$

$$\cdot \sum_{m=1}^{N}\left[\left(x_n^{(\mu_1)}\right)^{N-m} f_m\left(\vec{y}^{(\mu_1)}, d\vec{y}^{(\mu_1)}/dt\right)\right] , \tag{6.252}$$

with, in the right-hand side, the components of the N-vectors $\vec{y}^{(\mu_1)}(t)$ and $d\vec{y}^{(\mu_1)}(t)/dt$ replaced as follows by their expressions in terms of the components $x_n^{(\mu_1)}(t)$

respectively $\dot{x}_n^{(\mu_1)}(t)$ of the N-vectors $\vec{x}^{(\mu_1)}(t)$ respectively $d\vec{x}^{(\mu_1)}(t)/dt$ (see (6.241), (2.4) and (2.5)):

$$y_m^{(\mu_1)}(t) = (-1)^m \sum_{1 \le n_1 < n_2 < \cdots < n_m \le N} \left(x_{n_1}^{(\mu_1)} \cdot x_{n_2}^{(\mu_1)} \cdots x_{n_m}^{(\mu_1)} \right) , \qquad (6.253)$$

$$\dot{y}_1^{(\mu_1)}(t) = - \sum_{n=1}^{N} \left[\dot{x}_n^{(\mu_1)} \right] , \qquad (6.254a)$$

$$\dot{y}_m^{(\mu_1)}(t) = (-1)^m \sum_{j=1}^{m} \left\{ \dot{x}_j^{(\mu_1)} \sum_{1 \le n_1 < n_2 < \cdots n_{j-1}, n_{j+1}, \cdots < n_m \le N} \left[x_{n_1}^{(\mu_1)} \cdot x_{n_2}^{(\mu_1)} \cdots x_{n_m}^{(\mu_1)} \right] \right\} ,$$

$$m = 2, \ldots, N . \qquad (6.254b)$$

Note that in this manner we have identified $N!$ new N-body problems, labeled by the index μ_1 taking integer values in the range from 1 to $N!$ (with the significance of this index explained in the first part of this Chapter 6), and characterized by the $N!$ systems of N Newtonian equations of motion of goldfish type (6.252) (with (6.253) and (6.254)). These *new N-body problems* are *solvable*, since their solutions are provided by the *zeros* of the polynomials (6.251) which are themselves known because their *coefficients* are (a given permutation, characterized by the index μ_1, of) the solutions of the problem (6.249), which is assumed (to begin with) to be itself *solvable*.

It is now evident how this technique can be iterated over and over again in order to identify *new solvable N-body problems*. Let us just exhibit—relying on the notation of the first part of this Chapter 6—the $(N!)^2$ *solvable N-body problems* yielded by the next iteration of this procedure. The corresponding Newtonian equations of motion of goldfish type read as follows:

$$ddotx_n^{(\vec{\mu}^{(2)})} = \sum_{\ell=1, \, \ell \ne n}^{N} \left(\frac{2 \, \dot{x}_n^{(\vec{\mu}^{(2)})} \, \dot{x}_\ell^{(\vec{\mu}^{(2)})}}{x_n^{(\vec{\mu}^{(2)})} - x_\ell^{(\vec{\mu}^{(2)})}} \right)$$

$$- \left[\prod_{\ell=1, \, \ell \ne n}^{N} \left(x_n^{(\vec{\mu}^{(2)})} - x_\ell^{(\vec{\mu}^{(2)})} \right)^{-1} \right] \sum_{m=1}^{N} \left[\ddot{y}_m^{(\vec{\mu}^{(2)})} \left(x_n^{(\vec{\mu}^{(2)})} \right)^{N-m} \right] ,$$

$$(6.255a)$$

where $\vec{\mu}^{(2)}$ is the 2-vector with 2 components μ_1 and μ_2 (each of them taking integer values from 1 to $N!$). For each of the $(N!)^2$ assignments of this 2-vector

$\vec{\mu}^{(2)}$ (the significance of which has been explained in the first part of this Chapter 6), this system of N Newtonian equations of motion of goldfish type determines the time-evolution of the coordinates $x_n^{(\vec{\mu}^{(2)})}(t)$, with the quantity $\ddot{y}_m^{(\vec{\mu}^{(2)})}$ appearing in the right-hand side of (6.255a) being replaced by

$$\ddot{y}_m^{(\vec{\mu}^{(2)})} = \sum_{\ell=1,\ \ell \neq n}^{N} \left(\frac{2\,\dot{y}_m^{(\vec{\mu}^{(2)})}\,\dot{y}_\ell^{(\vec{\mu}^{(2)})}}{y_m^{(\vec{\mu}^{(2)})} - y_\ell^{(\vec{\mu}^{(2)})}} \right) - \left[\prod_{\ell=1,\ \ell \neq m}^{N} \left(y_m^{(\vec{\mu}^{(2)})} - y_\ell^{(\vec{\mu}^{(2)})} \right)^{-1} \right] \cdot$$
$$\cdot \sum_{n=1}^{N} \left[f_n \left(\vec{y}^{(\vec{\mu}^{(2)})},\ \dot{\vec{y}}^{(\vec{\mu}^{(2)})} \right) \left(y_m^{(\vec{\mu}^{(2)})} \right)^{N-n} \right], \tag{6.255b}$$

where moreover, in the right-hand side of this expression, the quantities $y_m^{(\vec{\mu}^{(2)})}(t)$ and $\dot{y}_m^{(\vec{\mu}^{(2)})}(t)$ should be replaced by their expressions in terms of the coordinates $x_n^{(\vec{\mu}^{(2)})}(t)$ and their time-derivatives $\dot{x}_n^{(\vec{\mu}^{(2)})}(t)$, as follows (see (2.4) and (2.5)):

$$y_m^{(\vec{\mu}_2)}(t) = (-1)^m \sum_{1 \leq n_1 < n_2 < \cdots < n_m \leq N} \left(x_{n_1}^{(\vec{\mu}_2)} \cdot x_{n_2}^{(\vec{\mu}_2)} \cdots x_{n_m}^{(\vec{\mu}_2)} \right), \tag{6.256}$$

$$\dot{y}_1^{(\vec{\mu}_2)}(t) = - \sum_{n=1}^{N} \left[\dot{x}_n^{(\vec{\mu}_2)} \right], \tag{6.257a}$$

$$\dot{y}_m^{(\vec{\mu}_2)}(t) = (-1)^m \sum_{j=1}^{m} \left\{ \dot{x}_j^{(\vec{\mu}_2)} \sum_{1 \leq n_1 < n_2 < \cdots n_{j-1}, n_{j+1}, \cdots < n_m \leq N} \left[x_{n_1}^{(\vec{\mu}_2)} \cdot x_{n_2}^{(\vec{\mu}_2)} \cdots x_{n_m}^{(\vec{\mu}_2)} \right] \right\},$$
$$m = 2, \ldots, N. \tag{6.257b}$$

As a particularly simple example, let us display the *solvable N*-body problems written above which obtain from the following simple assignment in the system of N decoupled ODEs (6.249):

$$f_n\left(\vec{x},\ \dot{\vec{x}}\right) = (\mathbf{i}\,\omega_n - \lambda_n)\dot{x}_n + \mathbf{i}\,\lambda_n\omega_n x_n, \tag{6.258a}$$

with the $2N$ parameters ω_n and λ_n *arbitrarily* assigned (but see also below later in this chapter for special assignments of these numbers). It is then clear—see, if need be, Subsection 4.3.1—that the solutions $x_n(t) = x_n^{(0)}(t)$ of the initial-value problems of the *seed* system of N ODEs (6.249) read as follows:

$$x_n^{(0)}(t) = (\mathbf{i}\,\omega_n + \lambda_n)^{-1} \left\{ \left[\lambda_n\,x_n^{(0)}(0) + \dot{x}_n^{(0)}(0) \right]\ \exp\left(\mathbf{i}\,\omega_n\,t\right) \right.$$
$$\left. + \left[\mathbf{i}\,\omega_n\,x_n^{(0)}(0) - \dot{x}_n^{(0)}(0) \right]\ \exp\left(-\lambda_n\,t\right) \right\}. \tag{6.258b}$$

The corresponding set of $N!$ *first-generation solvable* N-body problems is then characterized by the following Newtonian equations of motion of goldfish type (see (6.252))

$$\ddot{x}_n^{(\mu_1)} = \sum_{\ell=1,\,\ell\neq n}^{N} \left(\frac{2\,\dot{x}_n^{(\mu_1)}\,\dot{x}_\ell^{(\mu_1)}}{x_n^{(\mu_1)} - x_\ell^{(\mu_1)}} \right) - \left[\prod_{\ell=1,\,\ell\neq n}^{N} \left(x_n^{(\mu_1)} - x_\ell^{(\mu_1)} \right)^{-1} \right] \cdot$$
$$\cdot \sum_{m=1}^{N} \left(x_n^{(\mu_1)} \right)^{N-m} \left[(\mathbf{i}\,\omega_m - \lambda_m)\,\dot{y}_m^{(\mu_1)} + \mathbf{i}\,\omega_m\,\lambda_m\,y_m^{(\mu_1)} \right], \qquad (6.259)$$

with $y_m^{(\mu_1)}$ respectively $\dot{y}_m^{(\mu_1)}$ given in terms of the coordinates $x_\ell^{(\mu_1)}$ and $\dot{x}_\ell^{(\mu_1)}$ by (6.253) respectively (6.254).

Note that, in the special case in which the parameters ω_m and λ_m do not depend on the index m,

$$\omega_m = \omega\,, \quad \lambda_m = \lambda\,, \qquad\qquad\qquad (6.260a)$$

then via the *identities* (2.7) and (2.6), these equations of motion get simplified to read

$$\ddot{x}_n^{(\mu_1)} = \sum_{\ell=1,\,\ell\neq n}^{N} \left(\frac{2\,\dot{x}_n^{(\mu_1)}\,\dot{x}_\ell^{(\mu_1)}}{x_n^{(\mu_1)} - x_\ell^{(\mu_1)}} \right) + (\mathbf{i}\,\omega - \lambda)\,\dot{x}_n^{(\mu_1)}$$
$$+ \mathbf{i}\,\omega\,\lambda \left[\prod_{\ell=1,\,\ell\neq n}^{N} \left(x_n^{(\mu_1)} - x_\ell^{(\mu_1)} \right)^{-1} \right] \left(x_n^{(\mu_1)} \right)^{N}. \qquad (6.260b)$$

Note that, even though the *seed* model was *linear* and rather *trivially solvable*— see (6.249) with (6.258)—these (*solvable!*) Newtonian equations of motion are quite *nonlinear*.

And much more *nonlinear* are the equations of motion of the Newtonian equations of motion of goldfish type characterizing the *second generation* of *solvable* N-body problems, which read (see (6.255), and note that for simplicity we restrict attention to the case (6.260))

$$\ddot{x}_n^{(\vec{\mu}^{(2)})} = \sum_{\ell=1,\,\ell\neq n}^{N} \left(\frac{2\,\dot{x}_n^{(\vec{\mu}^{(2)})}\,\dot{x}_\ell^{(\vec{\mu}^{(2)})}}{x_n^{(\vec{\mu}^{(2)})} - x_\ell^{(\vec{\mu}^{(2)})}} \right)$$
$$- \left[\prod_{\ell=1,\,\ell\neq n}^{N} \left(x_n^{(\vec{\mu}^{(2)})} - x_\ell^{(\vec{\mu}^{(2)})} \right)^{-1} \right] \sum_{m=1}^{N} \left(x_n^{(\vec{\mu}^{(2)})} \right)^{N-m} \ddot{y}_m^{(\vec{\mu}^{(2)})},$$
$$(6.261a)$$

with the quantity $\ddot{y}_m^{(\vec{\mu}^{(2)})}$ appearing in the right-hand side of these equations of motion (6.261a) being replaced as follows:

$$
\ddot{y}_m^{(\vec{\mu}^{(2)})} = \sum_{\ell=1,\ \ell\neq n}^{N} \left(\frac{2\,\dot{y}_m^{(\vec{\mu}^{(2)})}\,\dot{y}_\ell^{(\vec{\mu}^{(2)})}}{y_m^{(\vec{\mu}^{(2)})} - y_\ell^{(\vec{\mu}^{(2)})}} \right) + (\mathbf{i}\,\omega - \lambda)\,\dot{y}_m^{(\vec{\mu}^{(2)})}
$$

$$
+\,\mathbf{i}\,\omega\,\lambda \left[\prod_{\ell=1,\ \ell\neq m}^{N} \left(y_m^{(\vec{\mu}^{(2)})} - y_\ell^{(\vec{\mu}^{(2)})} \right)^{-1} \right] \left(y_m^{(\vec{\mu}^{(2)})} \right)^{N}, \qquad (6.261b)
$$

with the additional replacements (6.256) and (6.257).

We leave as an easy exercise for the interested reader to ascertain the *necessary* and *sufficient* conditions on the parameters ω_m and λ_m which are *necessary and sufficient* to guarantee that the original N-body system of this example, see (6.249) with (6.258), has the property to be *isochronous* or *asymptotically isochronous* (for the notion of *asymptotic isochrony* see [54] or Chapter 6 of [27]; or see above Subsection 4.3.1). And it is clear that these features of the original N-body system of this example are then preserved by *all* the *solvable N*-body problems generated from this original model (of which the first two instances are exhibited above); this being due to the *inheritance* of the property of *isochrony* or *asymptotic isochrony*, as the case may be, by *all N*-body problems *generated* by a *seed* model possessing these features—as already noted above, see Section 2.3 and in particular, Remark 2.3.2.

Let us end this Section 6.1 with the following remark.

Remark 6.1.1. It is clear from our treatment that the endless sequences of *solvable N*-body problems of goldfish type associated via the technique described above to any *solvable seed* problem are in fact yielded by an appropriate sequence of *changes of dependent variables*. It might therefore be concluded that all these models are, as it were, *trivially equivalent* to the original *seed* model. But such an opinion would clash with the fact that *most*—perhaps *all*—the *solvable N*-body models which have been identified and investigated worldwide in the last few decades—their discovery and analysis constituting a substantial recent development of mathematics and mathematical physics—are as well reducible to *trivial time evolutions* by *appropriate changes of variables*. The rub is the identification—and the investigation—of the *appropriate changes of variables*; namely, in the present context, further study of the notion of generations of polynomials as described in Chapter 6, and more generally of the approach that has underlined all the findings reported in this book. ∎

6.2 The Peculiar Monic Polynomials The Zeros of Which Coincide with Their Coefficients

Over half a century ago S. M. Ulam [92] considered the well-known equivalence of the inversion of the transformation from N numbers x_n to their N symmetrical sums $\sigma_m(\tilde{x})$ multiplied by $(-1)^m$,

$$y_m = (-1)^m \, \sigma_m (\tilde{x}) \,, \quad m = 1, \ldots, N \,, \qquad (6.262\text{a})$$

$$\sigma_m (\tilde{x}) = \sum_{1 \le n_1 < n_2 < \ldots < n_m \le N} \left(x_{n_1} \cdot x_{n_2} \cdots x_{n_m} \right) \,, \qquad (6.262\text{b})$$

to the relations among the N *zeros* x_n and the N *coefficients* y_m of a monic polynomial of degree N (see (2.3)); and—while calling attention to this transformation (6.262) "operating on the N-dimensional real space or on the N-dimensional
complex space" [92]—he pointed out the interest of investigating its *fixed points*,
namely the identification of *all* the monic polynomials of degree N *with zeros equal
to their coefficients*. Soon afterwards this problem was completely solved in the
real domain—when the N polynomial *coefficients* and the N polynomial *zeros* are
all real numbers—by Paul R. Stein [90]. He proved that in this *real* context—
and restricting attention only to polynomials with *all* zeros hence *all* coefficients
nonvanishing (a well justified restriction, see below)—there are *no* polynomials
of this kind with degree $N > 4$. He moreover identified *all* the polynomials of
this kind with degree $N \le 4$: for $N = 2$ the single polynomial $(z - 1)(z + 2) =$
$z^2 + z - 2$ (already mentioned by Ulam [92]); for $N = 3$ the two polynomials
$(z - 1)(z + 1)^2 = z^3 + z^2 - z - 1$ and $z^3 + y_1 z^2 + y_2 z + y_3$, with $y_1 = y$, $y_2 = 1/y$,
$y_3 = 1/(y + 1)$ where $y \approx .56518772$ is the single *real* root of the cubic equation
$2y^3 + 2y^2 - 1 = 0$; for $N = 4$ the single polynomial $z^4 + y_1 z^3 + y_2 z^2 + y_3 z + y_4$ with
$y_1 = 1$, $y_2 = y \approx -1.75488$, $y_3 = 1/y$, $y_4 = -(y + 1)^2/y$, where y is identified
as the *single real* root of the cubic equation $y^3 + 2y^2 + y + 1 = 0$. [There is a
minor mistake in [90], where it is stated that y is the *single real* root in the range
$-2 < y < 0$ of the fourth-degree equation $y^4 + 3y^3 + 3y^2 + 2y + 1 = 0$, while
$y^4 + 3y^3 + 3y^2 + 2y + 1 \equiv (y + 1)(y^3 + 2y^2 + y + 1) = 0$ has of course $y = -1$ as
a second *real* root in the same range; but this second solution yields the polynomial
$z^4 + z^3 - z^2 - z = z(z - 1)(z + 1)^2$, a case that was excluded from consideration
in [90] because it features a *vanishing* coefficient and a *vanishing* root; see above
and below].

In Section 6.2 we report recent results that address in the more general context
of *complex* numbers the problem raised over half a century ago by Ulam [92]. A
motivation to do so is because this is generally a more natural context to discuss
properties of polynomials: see for instance the quite simple formulation of the
fundamental result stating that, in this *complex* context, the *number of zeros of a
polynomial equals its degree* (of course, taking into account the eventual *multiplicity* of some of these zeros). A second motivation is connected with the recently
introduced notion of *generations* of (monic) polynomials, characterized by the
property that the N *coefficients* of the polynomials of degree N of a *generation*
coincide with (one of the $N!$ permutations of) the N *zeros* of the polynomials of the
previous *generation*, as discussed above in this Chapter 6.

We now review—to help the reader who is only interested in the findings reported in this Section 6.2—the notation and terminology used below and we then report the main findings, which are then proven/justified in the last part of this Section 6.2.

Notation 6.2.1. Hereafter we suppose to work—unless otherwise indicated — with *complex* numbers and with *monic* polynomials of degree $N \geq 2$ in the *complex* variable z,

$$p_N \left(z; \vec{y}, \tilde{x} \right) = z^N + \sum_{m=1}^{N} y_m \, z^{N-m} = \prod_{n=1}^{N} \left(z - x_n \right) , \qquad (6.263)$$

which are characterized by their N *coefficients* y_m (the N components of the N-vector $\vec{y}$) and by their N *zeros* x_n (the N elements of the unordered set $\tilde{x}$). Above and hereafter indices such as n, m are *positive integers* in the range from 1 to N (unless otherwise indicated). The notation $p_N (z; \vec{y}, \tilde{x})$, see (6.263), is somewhat redundant, since this monic polynomial is equally well identified by assigning its N *coefficients* y_m *or* its N *zeros* x_n. Indeed its N coefficients y_m are *explicitly* expressed in terms of its N zeros by the well-known formulas (6.262). Conversely, the *unordered* set of the N zeros x_n of the polynomial (6.263) is uniquely identified when the N coefficients y_m of this polynomial are assigned, although *explicit* formulas expressing the *zeros* in terms of (square, cubic, respectively quartic roots of) the *coefficients* are generally only available for $N = 2$, $N = 3$, respectively $N = 4$. Let us finally note that occasionally below we will denote the N zeros and the N coefficients of the polynomial $p_N (z; \vec{y}, \tilde{x})$ with the more detailed notation $y_m^{(N)}$ and $x_n^{(N)}$, to avoid possible misunderstandings. ∎

Definition 6.2.1. The monic polynomial (6.263) is hereafter called *peculiar—"p-polynomial"*: $p_N^{(p)} (z)$—if it features the *peculiar* property that its N zeros x_n can be ordered so as to coincide one-by-one with its N coefficients y_m, implying

$$p_N^{(p)} \left(y_m; \vec{y}, \vec{y} \right) = 0 , \quad m = 1, \dots, N . \qquad (6.264)$$

Hereafter we use the notation $p_N^{(zp)} (z)$ to identify the subclass of these *p-polynomials* $p_N^{(p)} (z)$ which feature *at least one vanishing coefficient* hence *at least one vanishing zero*; and we note as an evident fact that *all* these polynomials $p_N^{(zp)} (z)$ are related to *all* the *p-polynomials* $p_{N-1}^{(p)} (z)$ by the neat formula

$$p_N^{(zp)} (z) = z \, p_{N-1}^{(p)} (z) , \qquad (6.265a)$$

which clearly also implies that *all* the *zp-polynomials* of degree N are given by the formula

$$p_N^{(zp)} (z) = z^k \, p_{N-k}^{(tp)} (z) , \quad k = 1, 2, \dots, N . \qquad (6.265b)$$

The polynomials appearing in the right-hand side of this formula belong to the sub-class of *p-polynomials* complementary to the subclass of *zp-polynomials*: they feature *no vanishing coefficient* hence *no vanishing zero*. They are hereafter called *truly pecu-liar polynomials—tp-polynomials*: $p_N^{(tp)}(z)$. We moreover indicate as $\nu_p(N)$, $\nu_{zp}(N)$ respectively $\nu_{tp}(N)$ the number of polynomials $p_N^{(p)}(z)$, $p_N^{(zp)}(z)$ respectively $p_N^{(tp)}(z)$ of degree N, with the relation $\nu_p(N) = \nu_{tp}(N) + \nu_{zp}(N)$ (valid counting multiplici-ties, see below). ∎

Let us now report the main findings via the following:

Proposition 6.2.1. (i) For *all* values of $N \geq 2$ the number $\nu_p(N)$ of *peculiar polynomials* $p_N^{(p)}(z)$ is, *counting multiplicities*, $\nu_p(N) = N!$. (ii) For all *generic* values of $N \geq 2$—i.e., for *all* values of N except for some *exceptional* values, see items (iii) and (iv) of this Proposition 6.2.1—the numbers $\nu_{zp}(N)$ of *zp-polynomials* $p_N^{(zp)}(z)$ and $\nu_{tp}(N)$ of *tp-polynomials* $p_N^{(tp)}(z)$ are, *counting multiplicities*, $\nu_{zp}(N) = (N-1)!$ and $\nu_{tp}(N) = N! - (N-1)! = (N-1)[(N-1)!]$. (iii) In the excep-tional $N = 4$ case, $\nu_{tp}(4) = 17 = 4! - 3! - 1$ and $\nu_{zp}(4) = 7 = 3! + 1$ (with a *zp-polynomial* counted *twice* because of its *multiplicity*: see below). (iv) It seems plausible to *conjecture* that there are *no* other exceptional cases; this is certainly true for $N < 7$. ∎

Note that, by definition, the rule $\nu_p(N) = \nu_{tp}(N) + \nu_{zp}(N)$ holds for all N; while for $N = 2, 3, 4, 5, 6$ the corresponding values of $\nu_p(N)$ respectively $\nu_{tp}(N)$ are $\nu_p(N) = 2, 6, 24, 120, 720$ respectively $\nu_{tp}(N) = 1, 4, 17, 96, 600$ (recall that, as reported above, in the *real* context the corresponding numbers are $\nu_{tp}(N) = 1, 2, 1, 0, 0 \ldots$, with $\nu_{tp}(N) = 0$ for $N > 4$; see [90]).

Remark 6.2.1. Let us reemphasize that, in counting these polynomials, account must be taken of their *multiplicity*, i.e., of the fact that some of them may coincide: in an analogous sense as in the formulation of the fundamental theorem of algebra stating that a polynomial of degree N features N *zeros* provided account is taken in counting the zeros of their *multiplicity*, in the exceptional cases of polynomials featuring *multiple zeros* (let us recall that a multiple zero x of order $j + 1$ of the polynomial $p_N(z)$ is characterized by the property $(d/dz)^k \, p_N(z) = 0$ at $z = x$ for $k = 0, 1, \ldots, j$, and $(d/dz)^k \, p_N(z) \neq 0$ at $z = x$ for $k = j + 1$, with k and j *nonnegative integers* and of course $j < N$). For a clarification in our context of the notion of *multiplicity* of *p-polynomials* see below, at the very end of this Section 6.2 (preferably after having digested what comes before that last paragraph...). ∎

Example 6.2.1. For $N = 2$ entailing $\nu_p(2) = 2$, $\nu_{tp}(2) = 1$, $\nu_{zp}(2) = 1$, the only *tp-polynomial* is $p_2^{(tp)}(z) = z^2 + z - 2 = (z - 1)(z + 2)$, and the only *zp-polynomial* is $p_2^{(zp)}(z) = z^2$. ∎

Example 6.2.2. For $N = 3$ entailing $\nu_p(3) = 6$, $\nu_{tp}(3) = 4$, $\nu_{zp}(3) = 2$, the 4 *tp-polynomials* $p_3^{(tp)}(z)$ are: (i) the polynomial $z^3 + z^2 - z - 1 = (z + 1)^2(z - 1)$ with two equal roots, and (ii) the 3 polynomials $z^3 + y_1 z^2 + y_2 z + y_3 = (z - y_1)(z - y_2)$

$(z - y_3)$ where $y_1 = y$, $y_2 = -1/y$, $y_3 = 1/(y+1)$ and the number y is one of the 3 roots of the cubic equation $2y^3 + 2y^2 - 1 = 0$ (only one of these roots is *real*, see above); while the 2 *zp-polynomials* $p_3^{(zp)}(z)$ are z^3 and $z^3 + z^2 - 2z = z(z-1)(z+2)$. ∎

Example 6.2.3. In the exceptional $N = 4$ case the 17 *tp-polynomials* $p_4^{(tp)}(z)$ are the following ones: (i) the 3 polynomials

$$z^4 + y_1 z^3 + y_2 z^2 + y_3 z + y_4 = (z-1)\ (z - y_2)\ (z - y_3)\ (z - y_4) \qquad (6.266)$$

with $y_1 = 1$, $y_2 = y$, $y_3 = -2 - 2y - y^2$, $y_4 = y(y+1)(y+2)$ where y is the root of the cubic equation $y^3 + 2y^2 + y + 1 = 0$ (featuring *one real* root and *two complex* roots); (ii) the 14 analogous polynomials (6.266) where instead

$$y_1 = y, \qquad (6.267a)$$

$$\begin{aligned} y_2 = \frac{1}{3301}(&-5780 - 9301\,y - 15701\,y^2 - 19444\,y^3 + 15074\,y^4 \\ &+ 62196\,y^5 + 79384\,y^6 + 62708\,y^7 + 7240\,y^8 - 87856\,y^9 \\ &- 157888\,y^{10} - 149344\,y^{11} - 79664\,y^{12} - 17888\,y^{13}), \end{aligned} \qquad (6.267b)$$

$$\begin{aligned} y_3 = \frac{1}{3301}\Big[&2950 - 7255\,y - 24398\,y^2 - 24629\,y^3 \\ &+ 30824\,y^4 + 51850\,y^5 + 60560\,y^6 + 34348\,y^7 \\ &- 35540\,y^8 - 129696\,y^9 - 116400\,y^{10} - 27936\,y^{11} \\ &+ 78384\,y^{12} + 69808\,y^{13} + 30528\,y^{14} - 9903\,y_2 \\ &- 9903\,(y_2)^2 - 3301\,(y_2)^3\Big], \end{aligned} \qquad (6.267c)$$

$$\begin{aligned} y_4 = \frac{1}{3301}\Big[&-2950 + 653\,y + 24398\,y^2 + 24629\,y^3 \\ &- 30824\,y^4 - 51850\,y^5 - 60560\,y^6 - 34348\,y^7 + 35540\,y^8 + 129696\,y^9 \\ &+ 116400\,y^{10} + 27936\,y^{11} - 78384\,y^{12} \\ &- 69808\,y^{13} - 30528\,y^{14} + 6602\,y_2 + 9903\,(y_2)^2 + 3301\,(y_2)^3\Big], \quad (6.267d) \end{aligned}$$

with y one of the 14 roots—*all different* among themselves and *complex*, constituted by 7 *complex conjugate* pairs—of the following *irreducible* polynomial equation of degree 14 (with *real* and *integer* coefficients):

$$\begin{aligned} 1 + 3\,y + 6\,y^2 + 6\,y^3 + 3\,y^4 - 12\,y^5 - 34\,y^6 - 44\,y^7 - 28\,y^8 \\ + 4\,y^9 + 48\,y^{10} + 80\,y^{11} + 80\,y^{12} + 48\,y^{13} + 16\,y^{14} = 0. \qquad (6.267e) \end{aligned}$$

The 7 *zp-polynomials* $p_4^{(zp)}(z)$ of degree 4 are: (i) the single polynomial z^4; (ii) the 2 polynomials $z^2\,p_2^{(tp)}(z)$, see above Example 6.2.1; and (iii) the 3 polynomials $z\,p_3^{(tp)}(z)$, see above Example 6.2.2. Altogether this amounts to 6 polynomials, but the polynomial $z\left(z^3 - z^2 + z - 1\right)$ arises *twice* (i.e., it has multiplicity 2: see below), so that in total we should count 7 *zp-polynomials* $p_4^{(zp)}(z)$ of degree 4. ∎

Example 6.2.4. For $N = 5$ it has also been possible—via the algebraic manipulation package *Mathematica* [82]—to obtain the $5! = 120$ *p-polynomials* $p_5^{(p)}(z)$. Of these 120 *p-polynomials* $p_5^{(p)}(z)$, $4! = 24$ are *zp-polynomials* $p_5^{(zp)}(z)$—of which only one, $z^5 + z^4 - z^3 - z^2$, with multiplicity 2—given in terms of the *p-polynomials* $p_4^{(p)}(z)$ by the rule (6.265b), while 96 are *tp-polynomials* $p_5^{(tp)}(z)$ *explicitly* identified in two groups: the first respectively the second group with their 5 *coefficients* and *zeros* identified by *explicit* formulas in terms of the 18 respectively 78 roots of two algebraic equations of order 18 respectively 78 *explicitly* given by *Mathematica* (the equations, not the roots). But these results are too complicated to be *explicitly* reported. ∎

In the remaining part of this Section 6.2, we report the proofs of the findings detailed above, and provide arguments to support the conjectures mentioned above.

The results listed above as Examples have been obtained or verified by *Mathematica* [82], and they can be thus re-obtained by the skeptical reader; this also holds for the nonexceptional nature of the finding for $N = 6$ (see item (iv) in Proposition 6.2.1). The only question that remains to be justified is the statement that the polynomial $p_4^{(zp)}(z) = z\left(z^3 - z^2 + z - 1\right) = z(z - 1)(z + 1)^2$ has multiplicity 2: see below the proof of Part (iii) of Proposition 6.2.1.

Part (i) of Proposition 6.2.1 is an immediate consequence of *Bézout Theorem* stating that a system of N *independent* algebraic equations of respective degrees k_m in N variables y_m ($m = 1, 2, \ldots, N$) features $\prod_{n=1}^{N}(k_n)$ solutions (counting the multiplicities of these solutions: see below). [For a proof of *Bézout Theorem* see any standard textbook of algebraic geometry, for instance [87] (see in particular Lesson VII therein); actually the main idea of this proof is outlined below, near the end of this Section 6.2. As a parenthetical remark, let us note here that, in applying Bézout's theorem, one should take into account *all* projective solutions, meaning that, apart from *finite* solutions, one should also consider solutions "at infinity" in the projective sense. It is however easy to see that such solutions do not exist in the present case, so that we may ignore this aspect of Bézout's theorem]. Clearly this *Theorem* implies Part (i) of Proposition 6.2.1 when applied to the system of N algebraic equations (of respective degree m)

$$y_m = (-1)^m \sum_{1 \leq n_1 < n_2 < \ldots < n_m \leq N} \left(y_{n_1} \cdot y_{n_2} \cdots y_{n_m}\right), \quad m = 1, \ldots, N, \qquad (6.268)$$

which are clearly implied—via the property $x_m = y_m$ characterizing *peculiar polynomials*, see (6.264)—by the N relations (6.262).

Part (ii) of Proposition 6.2.1 is an immediate consequence of Part (i) and of (6.265a).

To prove Part (iii) of Proposition 6.2.1, let us consider the 4 algebraic equations determining the 4 coefficients y_m of the *p-polynomials* $p_4^{(p)}(z)$ (see (6.268)):

$$y_1 = -(y_1 + y_2 + y_3 + y_4) , \tag{6.269a}$$

$$y_2 = y_1 y_2 + y_1 y_3 + y_1 y_4 + y_2 y_3 + y_2 y_4 + y_3 y_4 , \tag{6.269b}$$

$$y_3 = -(y_1 y_2 y_3 + y_1 y_2 y_4 + y_1 y_3 y_4 + y_2 y_3 y_4) , \tag{6.269c}$$

$$y_4 - y_1 y_2 y_3 y_4 . \tag{6.269d}$$

Now Bézout's Theorem applied to this system of 4 algebraic equations of respective degrees 1, 2, 3, 4 tells us that there are $4! = 24$ solutions of these 4 algebraic equations in the 4 unknowns y_1, y_2, y_3, y_4, confirming what was already stated in Part (i) of Proposition 6.2.1, i.e. $\nu_p(4) = 24$.

Let us then try and ascertain the number $\nu_{tp}(4)$ of *tp-polynomials* $p_4^{(tp)}(z)$. For these polynomials none of the 4 coefficients y_m should vanish, hence we can assume that

$$y_4 \neq 0 . \tag{6.270}$$

It is then justified to replace the last one of the 4 equations (6.269) with the simpler equation (of degree 3 rather than 4)

$$1 = y_1 y_2 y_3 . \tag{6.271}$$

There obtains thereby a new system of 4 algebraic equations in the 4 unknowns y_1, y_2, y_3, y_4 which has, by Bézout's Theorem, $1 \cdot 2 \cdot 3 \cdot 3 = 18$ solutions. It is indeed possible, via *Mathematica* [82], to find the 17 solutions of this system listed above (see Example 6.2.3) and in addition the solution

$$y_1 = 1 , \quad y_2 = y_3 = -1 , \quad y_4 = 0 , \tag{6.272}$$

which is indeed easily seen to satisfy the first 3 of the 4 equations (6.269) as well as (6.271) (and it clearly satisfies trivially also the last of the 4 equations (6.269)). But, contrary to our assumption in (6.270), this solution features a vanishing y_4; hence the corresponding polynomial $z^4 + z^3 - z^2 - z = z(z-1)(z+1)^2$ is a *zp-polynomial* $p_4^{(zp)}(z)$ rather than a *tp-polynomial* $p_4^{(tp)}(z)$, explaining why there are only 17, rather than 18, *tp-polynomials* $p_4^{(tp)}(z)$, i.e., $\nu_{tp}(4) = 17$.

To confirm the proof of Part (iii) of Proposition 6.2.1, let us evaluate in an analogous manner the number $\nu_{zp}(4)$ of *zp-polynomials* of degree 4. These polynomials $p_4^{(zp)}(z)$ are characterized by the complementary condition to (6.270), namely

$y_4 = 0$, implying that the system of 4 algebraic equations (6.269) reduces to the following system of 3 algebraic equations in the 3 variables y_1, y_2, y_3:

$$y_1 = -(y_1 + y_2 + y_3) , \tag{6.273a}$$

$$y_2 = y_1\, y_2 + y_1\, y_3 + y_2\, y_3 , \tag{6.273b}$$

$$y_3 = -y_1\, y_2\, y_3 . \tag{6.273c}$$

By Bézout's Theorem, this system has $3! = 6$ solutions (which are those already obtained above: see Example 6.2.2), and it is easily seen that one of these 6 solutions just reproduces the solution (6.272) of the system (6.269); the solution thus obtained is a simple solution of (6.273) as well as a simple solution of the system of equations defined by (6.269a, 6.269b, 6.269c, 6.271), and hence we attribute to this solution multiplicity 2, implying $\nu_{zp} = 6 + 1 = 7$ as stated under Part (iii) of Proposition 6.2.1, which is thereby validated.

Finally, to justify the conjectural part mentioned in Part (iv) of Proposition 6.2.1—the nonconjectural part being confirmed by a direct computation via *Mathematica* [82]—let us examine in more detail the special $N = 4$ case with $\nu_{zp}(N) \neq (N-1)!$ and $\nu_{tp}(N) \neq N! - (N-1)!$. This outcome obtained for $N = 4$ because in that case we started from the assumption (6.270), which allowed us to reduce the investigation of the *truly peculiar* polynomials $p_4^{(tp)}(z)$ of degree 4, rather than to the complete set of equations (6.269), to the somewhat simpler set with the last of these 4 algebraic equations replaced by the simplified algebraic equation (6.271) (of degree 3 rather than 4); but we then discovered that one of the solutions of this simplified set of equations violated the assumed inequality (6.270), hence yielded as solution a *zp-polynomial* rather than a *tp-polynomial*. Thus, to justify the conjectural part of Part (iv) of Proposition 6.2.1, we must try and provide arguments to back our expectation that an analogous phenomenon does *not* happen for $N > 4$. The following intermediate result is instrumental to that end.

Let us consider for arbitrary N the system of algebraic equations (6.268) together with the assumption

$$y_N \neq 0 , \tag{6.274a}$$

which allows to replace the last equation of the system (6.268), i.e., $\prod_{m=1}^{N} y_m = (-1)^N y_N$, with the simplified equation (of degree $N - 1$ rather than N) reading

$$\prod_{\ell=1}^{N-1} y_\ell = (-1)^N . \tag{6.274b}$$

Let us then consider—in the spirit of an argument *per absurdum*, see below—that the resulting system of N algebraic equations features a solution which does violate the inequality (6.274a), i.e., that it entails a *vanishing* value for the quantity y_N, $y_N = 0$. It is then easily seen that the first $N - 1$ algebraic equations of the system under consideration, with this assignment, reproduce just the system for the investigation of *peculiar polynomials* of degree $N - 1$, i.e., they read as follows (with $y_\ell^{(N-1)}$ indicating the $N - 1$ *coefficients* and *zeros* of the polynomial $p_{N-1}^{(tp)}(z)$):

$$y_\ell^{(N-1)} = (-1)^\ell \sum_{1 \le n_1 < n_2 < \ldots < n_\ell \le N-1} y_{n_1}^{(N-1)} \cdot y_{n_2}^{(N-1)} \cdots y_{n_\ell}^{(N-1)} , \quad \ell = 1, \ldots, N-1 ,$$

(6.275a)

while the additional relation (6.274b), together with the last of these equalities (6.275a) which reads

$$y_{N-1}^{(N-1)} = (-1)^{N-1} \prod_{\ell=1}^{N-1} \left[y_\ell^{(N-1)} \right] , \tag{6.275b}$$

entail the additional condition

$$y_{N-1}^{(N-1)} = -1 . \tag{6.276}$$

We thus conclude that the system of equations relevant for the *tp-polynomials* of degree N can only give the *anomalous* result—a solution with $y_N^{(N)} = 0$, hence to be categorized as a *zp-polynomial* polynomial rather than a *tp-polynomial*—if at least one of the $N - 1$ *tp-polynomials* of degree $N - 1$ feature as its last coefficient $y_{N-1}^{(N-1)}$ the value -1, see (6.276).

That the *anomalous* phenomenon—one or more of the $(N - 1)\,!$ values of the coefficient $y_{N-1}^{(N-1)}$ equal -1, see (6.276)—does not happen for $N < 4$ nor for $N = 5$, while it does happen for $N = 4$—is shown by the Examples reported above (in this Section 6.2). To proof the conjectural component of Part (ii) of Proposition 6.2.1, it would therefore be sufficient to prove that no *tp-polynomial* $p_N^{(tp)}(z)$ with $N > 4$ feature a *last* coefficient $y_N^{(N)} = -1$. We now try and demonstrate this result, but at the end we indicate why this proof is not quite cogent—although we feel it provides a justification for our proffering the *conjecture* mentioned in Part (iv) of Proposition 6.1.

Let us indeed apply the standard "reduction" mechanism—that is actually instrumental to prove Bézout's Theorem—to the system of N algebraic equations characterizing the *coefficients* and *zeros* y_m of the polynomial $p_N^{(tp)}(z)$ with $N > 4$ and $y_N^{(N)} \ne 0$, reading as follows:

$$y_m^{(N)} = (-1)^m \sum_{1 \le n_1 < n_2 < \ldots < n_m \le N} \left(y_{n_1}^{(N)} \cdot y_{n_2}^{(N)} \cdots y_{n_m}^{(N)} \right) , \quad m = 1, \ldots, N-1 , \tag{6.277a}$$

$$y_1^{(N)} \, y_2^{(N)} \cdots y_{N-1}^{(N)} = (-1)^N . \tag{6.277b}$$

This standard reduction technique amounts to the systematic iterative application of the observation according to which any polynomial equation

$$P_M(z) = \sum_{m=0}^{M} \left(c_m \, z^{M-m} \right) = 0 \, , \tag{6.278a}$$

of *arbitrary positive integer* degree M—hence with $c_0 \neq 0$—implies for *all* the values z_m of its M roots the formula

$$(z_m)^M = -(c_0)^{-1} \sum_{\ell=1}^{M} \left[c_\ell \, (z_m)^{M-\ell} \right] \, . \tag{6.278b}$$

This formula allows a systematic replacement of the *highest* power of any one of the roots z_m of the polynomial equation (6.278a) with an appropriate combination of *lesser* powers of that root. This procedure—which can be *explicitly* performed, step by step, only for relatively small values of N, becoming unmanageably complicated for large N—allows nevertheless to demonstrate that the system (6.277) leads to the following system of N *independent* nonlinear algebraic formulas:

$$y_\ell^{(N)} = R_\ell \left(y_N^{(N)} \right) \, , \quad \ell = 1, 2, \dots, N-1 \, , \tag{6.279a}$$

$$P_{\nu_{tp}(N)} \left(y_N^{(N)} \right) = 0 \, , \tag{6.279b}$$

where the $N-1$ functions $R_\ell(z)$ with $\ell = 1, 2, 3, \dots, N-1$ are *rational* functions of their argument z with *real* (in fact *rational*) *coefficients*, while $P_{\nu_{tp}(N)}(z)$ is a polynomial in z of (*even*) degree $\nu_{tp}(N) = (N-1) \, [(N-1)!]$ (see above) with *real*, in fact *integer*, *coefficients*. Hence the $\nu_{tp}(N)$ solutions of this system—identifying the $\nu_{tp}(N)$ sets of N-vectors $\vec{y}^{(N)}$, the components of which identify the N *coefficients* and *zeros* $y_m^{(N)}$ of the N *tp-polynomials* $p_N^{(tp)}(z)$—are provided by the $\nu_{tp}(N)$ values of the *zeros* $y_N^{(N)} = z_k$ (with $k = 1, \dots, \nu_{tp}(N)$) of the polynomial equation (6.279b)—each of which then identifies univocally the remaining $N-1$ coefficients and zeros $y_\ell^{(N)}$ (with $\ell = 1, 2, 3, \dots, N-1$) of a polynomial $p_N^{(tp)}(z)$ via the $N-1$ formulas (6.279a).

> **Remark 6.2.2.** This argument also characterizes the possible occurrence of *multiple tp-polynomials* $p_N^{(tp)}(z)$ as corresponding to the possible occurrence of *multiple zeros* of the polynomial $P_{\nu_{tp}(N)}(z)$, see (6.279b). ∎

Our remaining task is to exclude the possibility—for all values of $N \geq 4$—that $y_N^{(N)} = -1$, i.e., that the polynomial equation (of *even* degree $\nu_N^{(tp)}$) $P_{\nu_{tp}(N)}(z) = 0$, see (6.279b), feature the zero $z = -1$. Suppose *per absurdum* that this is instead the case, namely that $z = -1$ is a zero of the polynomial $P_{\nu_{tp}(N)}(z)$, i.e., $P_{\nu_{tp}(N)}(-1) = 0$ for some value of $N \geq 4$. Then the formulas (6.279a) would imply that *all* the corresponding values $y_n^{(N)}$ (with $n = 1, \dots, N-1$) would also be *real* numbers. There would therefore exist, for $N \geq 5$, a *real tp-polynomial* $p_N^{(tp)}(z)$. But the impossibility of this outcome has been demonstrated by Paul Stein half a century ago [90]. The conclusion, therefore, seems to follow that for all $N \geq 4$, $P_{\nu_{tp}(N)}(-1) \neq 0$ (in fact, more generally, that the *real* polynomial $P_{\nu_{tp}(N)}(z)$ of *even* degree $\nu_N^{(tp)}$ cannot feature *any real zero*).

But unfortunately this argument has a weak point: it might happen that at some step of the reduction process one of the equations involved in this process turns out *not to be* linear in the quantity to be computed, thereby possibly causing its solutions not to be necessarily *real*; thereby causing all the subsequent steps of the reduction process to involve *complex* numbers, invalidating the concluding part of the above purported proof. Hence the statement in Part (iv) of Proposition 6.1 maintains its *conjectural* status, which might however be strengthened by these argumentations.

Let us end this Section 6.2 with a final survey of the notion of *multiplicity* of *peculiar* polynomials, to emphasize that in our treatment there are *two distinct* origins possibly accounting for this phenomenon. The *first* is associated with the identification of *peculiar* polynomials via the solution of systems of algebraic equations characterizing their coefficients, and the related utilization of Bézout's Theorem to identify and count such polynomials. In this context the notion of the eventual *multiplicity* of a *peculiar* polynomial is directly related to the *multiplicity* of the root of the single high-order algebraic equation to which the identification of the *peculiar* polynomial is traced (as outlined above). A *second* notion—relevant to the subclasses of *zp-polynomials* and *tp-polynomials*—is clearly exemplified by the fact that the class of algebraic equations characterizing the *tp-polynomials*—simplified thanks to the assumption that $y_N \neq 0$ (note in the $N = 4$ case discussed above the replacement of the algebraic equation (6.269d) of degree 4 with the algebraic equation (6.271) of degree 3)—might then turn out to yield a solution with $y_N = 0$ (as it indeed happens in the exceptional $N = 4$ case), which therefore identifies a *zp-polynomial* rather than a *tp-polynomial*, thereby causing a decrease of the number of *tp-polynomials* and a corresponding increase in the number of *zp-polynomials* (which generally then causes—according to our interpretation—an increase in the multiplicity of that *zp-polynomial*).

6.N Notes on Chapter 6

For the results reported in the first part of Chapter 6 and in Section 6.1, see [5].

For the results described in Section 6.2, see [67], where the results reported are actually somewhat more general than those described above, because they also discuss—in addition to the special subclass of *peculiar* polynomials characterized by the property to feature (at least) one *coefficient* (and *zero*) having the value 0—the special subclass of *peculiar* polynomials characterized by the property to feature (at least) one *coefficient* (and *zero*) having the value 1. There is therefore a *difference* among the definition of *truly peculiar* polynomials given in [67] and that given above (this definition—see Definitions 6.2.1—is closer to that originally implied by Ulam [92] and Stein [90], although neither one of them used the terminology "*peculiar* polynomials" or "*truly peculiar* polynomials").

Let us end this brief Section 6.N with the mention of an interesting *open* problem which might be considered a natural follow-up to the results reported above in the first part of this Chapter 6 and in its Section 6.2. Indeed, in Section 6.2 findings are reported—in the context of *complex* numbers—on the question raised over half a century ago by Ulam [92] and completely solved—in the context of *real* numbers—by Stein [90]. This issue consists of the determination of the *fixed points* of the relation among the N coefficients and the N zeros of a monic polynomial of degree N. The notion of *generations* of polynomials reported in the first part of this Chapter 6 suggests an analogous question about the *fixed points* of the relation among the N coefficients of a monic polynomial of degree N belonging to the *k-th generation*, to the N coefficients of the monic polynomial of degree N which plays the role of *seed* for the relevant *generations* of polynomials (with $k > 1$; the case with $k = 1$, corresponding to the problem originally raised by Ulam and discussed in Section 6.2).

Finally, let us note that, to the best of our knowledge, after the original papers by Ulam [92] and Stein [90], no other papers have been issued on the problem discussed in Section 6.2, except for a paper by A. J. Di Scala and O. Maciá in 2008 [71] and one by O. Bihun and D. Folghesu in 2018 [9] (beside the paper [67] on which all the results reported above are based).

7

Discrete Time

In Chapter 7 a technique is introduced and tersely investigated that allows to generate—starting from any *solvable discrete-time* dynamical system involving N time-dependent variables—an endless hierarchy featuring, at its k-th level (with $k = 1, 2, 3, \ldots$), $(N!)^k$ *new*, generally *nonlinear, discrete-time* dynamical systems, also involving N time-dependent variables and being *solvable* by *algebraic* operations (i.e., essentially by finding the N *zeros* of explicitly known monic polynomials of degree N). These dynamical systems may also feature large numbers of *arbitrary* constants, and they need *not* be *autonomous*. The *solvable* character of these models allows to identify special cases with remarkable time evolutions: for instance, *isochronous* or *asymptotically isochronous* dynamical systems. A few examples are reported.

The models under consideration in this paper describe the evolution in the *discrete-time* variable $\ell = 0, \pm 1, \pm 2, \ldots$ of an arbitrary number N of identical points moving in the *complex* x-plane, the positions of which are characterized by N *complex* coordinates—for instance $x_n \equiv x_n(\ell)$ or $y_n(\ell)$. Both the equations of motion characterizing these models, and their solutions involve only the *algebraic* operation of finding the N zeros of an explicitly known ℓ-dependent polynomial of degree N in z (and in addition, for initial-value problems, of solving the *explicitly solvable* algebraic systems of a finite number of *linear* equations). The technique to identify such models is analogous to, but somewhat more flexible than, the approach employed in [29], [31], [64], [65], [66], and [13] to identify and discuss several *solvable discrete-time* many-body problems.

> **Notation 7.1.** Unless otherwise indicated, indices such as n, m, j run over the integers from 1 to N, with N a given positive integer ($N \geq 2$). All quantities—except those taking *integer* values, such as the indices and the discrete-time ℓ—are *complex* numbers. Hereafter superimposed arrows denote N-vectors, for instance the N-vector $\vec{y} \equiv (y_1, y_2, \ldots, y_N)$ features the N components y_n, while the notation $\tilde{x} \equiv \{x_1, x_2, \ldots x_N\}$ denotes the *unordered* set of N numbers x_n. In the following we will generally limit consideration to the *generic* case in which the N complex

numbers x_n are *all different among themselves*. And as usual *empty* sums vanish, and *empty* products equal *unity* (see Section 2.1).

In the following—as in Chapter 6—it will be important to associate to an *unordered* set $\tilde{x}$ the N-vector the components of which correspond to a specific ordering assignment of its N elements x_n. There are generally $N!$ different such vectors, and we will use for them the notation $\vec{x}_{[\mu]} \equiv \left(x_{[\mu],1}, x_{[\mu],2}, \ldots, x_{[\mu],N}\right)$, with the index μ identifying a specific ordering assignment of the N numbers x_n—hence taking the $N!$ *integer* values from 1 to $N!$ (since N different objects can be ordered in $N!$ different ways). ∎

Remark 7.1. In the following we mainly focus on dynamical systems characterized by *first-order* equations of motion, say

$$y_m\left(\ell + 1\right) = f_m\left(\vec{y}\left(\ell\right);\, \ell\right)\; ; \tag{7.280}$$

but occasionally below it will also be of interest to consider higher-order equations of motion. ∎

Remark 7.2. The *solvable* equations of motion identified below determine the *discrete-time* motions in the *complex* x-plane of N points characterized by the N coordinates $x_n \equiv x_n(\ell)$, defined as an *unordered* set $\tilde{x}(\ell) \equiv \{x_1(\ell), x_2(\ell), \ldots, x_N(\ell)\}$, indeed generally as the N zeros of a ℓ-dependent polynomial $p_N(z; \ell)$ of degree N in z. Hence these equations of motion are only deterministic inasmuch as they identify (uniquely) the *unordered* set $\tilde{x}(\ell + 1)$ in terms of the *unordered* set $\tilde{x}(\ell)$, but they do *not* associate each coordinate $x_n(\ell)$ to a specific value of the index n labeling it. So these models describe N *indistinguishable* point particles, the positions of which in the *complex* x-plane at the discrete-time ℓ are characterized by the N coordinates $x_n(\ell)$. A *preferred* association of these coordinates to their labels might be provided at each step of the *discrete-time* evolution by an argument of *contiguity*, but this would be relevant only if the N values $x_n(\ell + 1)$ are *all* adequately *separated from each other*—in the complex x-plane—and each of them is *adequately close to one and only one* of the N values $x_m(\ell)$ being themselves *well separated among each other*; more about this below. ∎

The main idea of previous models (see [29], [31], [64], [65]) was to identify a solvable *discrete-time* evolution of an $N \times N$ matrix and to then focus on the evolution of its N eigenvalues. A more straightforward approach—already employed in [66] and [13]—focused on the *solvable* evolution of a polynomial $p_N(z; \ell)$ of degree N in z and on that of its N zeros $x_n(\ell)$. The more convenient approach employed here—consistently with the rest of this book—focuses more directly on the *discrete-time* evolutions of the N coefficients $y_m(\ell)$ respectively the N zeros $x_n(\ell)$ of a time-dependent polynomial $p_N(z; \ell)$ by taking advantage of a key formula relating these evolutions, see below. The *solvable discrete-time* dynamical systems that can be identified via this approach are described in the following Section

7.1, a few examples are discussed in this Section 7.2, while in Section 7.3—after repeating an important consideration—we tersely outline a possible development of the results reported in Chapter 7.

7.1 Solvable Discrete-Time Dynamical Systems

A main protagonist of our treatment is the time-dependent monic polynomial (analogous to (2.3), but note the replacement of the *continuous* time t with the *discrete* time ℓ):

$$p_N(z;\ell) \equiv p_N(z;\vec{y}(\ell);\tilde{x}(\ell)) = z^N + \sum_{m=1}^{N}\left[y_m(\ell)\,z^{N-m}\right] = \prod_{n=1}^{N}\left[z - x_n(\ell)\right].$$

$$(7.281a)$$

Remark 7.1.1. The notation $p_N(z;\vec{y}(\ell);\tilde{x}(\ell))$ is *redundant*, since to define this monic polynomial of degree N in the variable z at any time ℓ it is clearly sufficient to assign *either* its N coefficients $y_m(\ell)$ *or* its N zeros $x_n(\ell)$. Indeed, the N coefficients $y_m(\ell)$ are defined in terms of the N zeros $x_n(\ell)$ by the standard formulas (2.4), which now read as follows:

$$y_m(\ell) = (-1)^m\,\sigma_m\,(\tilde{x}(\ell))$$
$$= (-1)^m \sum_{1 \leq n_1 \leq n_2 \leq \cdots \leq n_m}\left[x_{n_1}(\ell)\,x_{n_2}(\ell)\cdots x_{n_m}(\ell)\right];$$

$$(7.281b)$$

and conversely the N zeros $x_n(\ell)$ are uniquely determined (but only as an *unordered* set) by the N coefficients $y_m(\ell)$, although *explicit* formulas to this effect are only available for $N \leq 4$. ∎

The main tool of our approach is the following key *identity*, implied by (7.281a) and proven at the end of this Section 7.1:

$$\prod_{j=1}^{N}\left[x_n(\ell+p) - x_j(\ell)\right] + \sum_{m=1}^{N}\left\{\left[y_m(\ell+p) - y_m(\ell)\right]\left[x_n(\ell+p)\right]^{N-m}\right\} = 0,$$

$$(7.282)$$

which holds for *all integer* values of p. Note that this formula entails that the N values of the variables $x_n(\ell+p)$—for the N values of the index n in the range from 1 to N—are the N zeros of the following polynomial of degree N in the *complex* variable z:

$$\hat{p}_N(z;\ell) = \prod_{j=1}^{N}\left[z - x_j(\ell)\right] + \sum_{m=1}^{N}\left\{\left[y_m(\ell+p) - y_m(\ell)\right](z)^{N-m}\right\},$$

$$(7.283a)$$

$$\hat{p}_N\left(z;\,\ell\right) = \prod_{j=1}^{N}\left[z - x_j\left(\ell + p\right)\right].\tag{7.283b}$$

The merit of this formula—in either one of its equivalent versions (7.282) or (7.283)—is to relate the *discrete-time* evolution of the *N zeros* $x_n(\ell)$ to the *discrete-time* evolution of the *N coefficients* $y_m(\ell)$, allowing to identify directly the equations of motion of *new solvable* dynamical systems from those of *known solvable* dynamical systems. Indeed let us assume that the *N* quantities $y_m(\ell)$ evolve in the *discrete-time* variable ℓ according to the following dynamical system:

$$y_m\left(\ell + p\right) = f_m\left(\vec{y}(\ell),\,\vec{y}\left(\ell + 1\right),\ldots,\,\vec{y}\left(\ell + p - 1\right);\,\ell\right),\tag{7.284}$$

where the *N* functions f_m are conveniently assigned so that this system is *solvable*. For instance this *solvable* dynamical system might be any one of those treated in the papers [29], [31], [64], [65], [66], [13]. Note that, at this stage, we do not exclude that the functions f_m might feature an explicit dependence on the discrete-time variable ℓ, implying thereby that the corresponding *solvable* dynamical system (7.284) is *not* autonomous; there indeed exist *nonautonomous discrete-time* dynamical systems that are nevertheless *solvable*, for instance the nonlinear system treated in [66] or the rather trivial *decoupled linear* system (with $p = 1$)

$$y_m\left(\ell + 1\right) = g_m(\ell)\,y_m(\ell) + h_m(\ell)\tag{7.285a}$$

with $g_m(\ell)$ and $h_m(\ell)$ arbitrarily assigned functions, the solution of which reads as follows (see Notation 7.1):

$$y_m(\ell) = y_m\left(0\right)\prod_{\ell'=0}^{\ell-1}\left[g_m\left(\ell'\right)\right] + \sum_{\ell''=0}^{\ell-1}\left\{\left[\prod_{\ell'=\ell''+1}^{\ell-1} g_m\left(\ell'\right)\right]h_m\left(\ell''\right)\right\}.\tag{7.285b}$$

It is then clear that the dynamical system characterized by the following equations of motion (implied by (7.282) and (7.284))

$$\prod_{j=1}^{N}\left[x_n\left(\ell + p\right) - x_j(\ell)\right]$$

$$+ \sum_{m=1}^{N}\left\{\left[f_m\left(\vec{y}(\ell),\ldots,\vec{y}\left(\ell + p - 1\right);\ell\right) - y_m(\ell)\right]\left[x_n\left(\ell + p\right)\right]^{N-m}\right\} = 0$$

$$\tag{7.286a}$$

is as well *solvable*. Note that (7.286a) is equivalent to the prescription that the *N* updated values $x_n\left(\ell + p\right)$ of the *N* coordinates $x_n(\ell)$ coincide with the *N* zeros of the following polynomial equation of degree *N* in z,

$$\prod_{j=1}^{N}\left[z - x_j(\ell)\right] + \sum_{m=1}^{N}\left\{\left[f_m\left(\vec{y}(\ell),\ldots,\,\vec{y}\left(\ell + p - 1\right);\,\ell\right) - y_m(\ell)\right]z^{N-m}\right\} = 0.$$

$$\tag{7.286b}$$

In both of the last two formulas the N components $y_m(j)$ of the N-vector $\vec{y}(j)$, where $j = \ell, \ell+1, \ldots, \ell+p-1$, should be replaced by their expressions (7.281b) in terms of the N components of the unordered set $\tilde{x}(j)$ with $j = \ell, \ell+1, \ldots, \ell+p-1$.

The claim that this dynamical system (7.286) is *solvable* is validated by the fact that the coordinates $x_n(\ell)$, yielding its solution at time ℓ, are the N *zeros* of the polynomial (7.281a), of degree N in z—hence they are obtainable by an *algebraic* operation—while the *coefficients* $y_m(\ell)$ of this polynomial (7.281) are themselves obtainable by *algebraic* operations, since the dynamical system (7.284) characterizing the time-evolution of these quantities is by assumption itself *solvable*. In particular the *initial-value* problem for this dynamical system (7.286) can be solved—for an arbitrary assignment of the "initial" data $x_n(j)$ with $j = 0, 1, \ldots, p-1$—via the following 3 steps: (i) compute the corresponding N *initial* data $y_m(j)$ with $j = 0, 1, \ldots, p-1$ of the dynamical system (7.284) via (7.281b); (ii) obtain the N values $y_m(\ell)$ for $\ell \geq p$ via the *solvable* dynamical system (7.284) with the initial data computed in step (i); (iii) find the N zeros $x_n(\ell)$ with $\ell \geq p$ of the polynomial $p_N(z; \ell)$, see (7.281a), defined by its N *coefficients* $y_m(\ell)$ as computed in step (ii).

Specific examples of *solvable* dynamical systems (7.286) will be exhibited and discussed in the following Section 7.2.

The usefulness of the key formula (7.282) is not limited to the identification of the new *solvable* dynamical system (7.286) associated to the previously known *solvable* dynamical system (7.284): it also opens the possibility to *iterate*—albeit up to a limitation that we explain below—the procedure we used above in order to obtain the *new* solvable dynamical system (7.286) from the *known solvable* system (7.284), by then using as known input in this approach just the newly identified *solvable* dynamical system (7.286). And clearly this approach can be iterated over and over again, yielding endless hierarchies of *new solvable* dynamical systems, in analogy to the treatment for *continuous-time* (see Chapter 6).

> **Remark 7.1.2.** There is however a significant difference with respect to the analogous treatment in the *continuous-time* case, see the preceding chapters of this book. The difference originates from the fact that in the *continuous-time* case the time evolution is in fact completely *deterministic*, implying that in that context one is dealing with *distinguishable* particles: the assignment of the initial values $x_n(0)$ of the particle positions in the *complex* x-plane at the initial time $t = 0$ determines *unambiguously*—thanks to the *continuity* of the time-evolution of the particle coordinates $x_n(t)$ as functions of the *continuous-time* variable t—the particle positions $x_n(t)$ for all future time t: say, the coordinate $x_1(t)$ is the one that has evolved continuously over time from the initial datum $x_1(0)$, and likewise for all values of the labels n in the range from 1 to N identifying the coordinates $x_n(t)$. Note that this is the case even though the particle positions $x_n(t)$ are, also in that context, identified with the *zeros* of a time-dependent polynomial $p_N(z; t)$ of degree N in the *complex* variable z.

In the *discrete-time* case treated in Chapter 7 one is instead dealing—as explained above—with *indistinguishable* particles. This does not cause any problem for the definition and solution of the *discrete-time* model (7.286) described just above, which characterizes the evolution in the *discrete-time* variable ℓ of the *unordered* set of coordinates $\tilde{x}(\ell)$. But it raises an issue when we try to use that model to describe the evolution of the *coefficients* $y_m(\ell)$ of a polynomial, since this set of numbers is an *ordered* set. ∎

To further explain, in the simplest setting, what we mean—and thereby also clarify some relevant differences among the *continuous-time* case treated in the preceding chapters of this book and the *discrete-time* case treated in this Chapter 7—let us hereafter focus on equations of motion of *first* order, in particular let us refer—in the remaining part of this Section 7.1—to the $p = 1$ special case of the dynamical systems (7.284) and (7.286) and of the key formula (7.282). The interested reader will have no difficulty to extend the following treatment to values of the *integer* parameter p larger than *unity*.

In the following we will refer to the *solvable* dynamical system (7.284) (with $p = 1$) as the *seed dynamical system* and to the *new* dynamical system (7.286) (with $p = 1$)—the *solvable* character of which has been detailed just above—as the *generation zero dynamical system*. We will also refer to the polynomial (7.281a) as the *generation zero* polynomial.

Let us then try and iterate the process used to obtain the *generation zero dynamical system* (7.286) from the *seed system* (7.284). The idea—following the analogous treatment in the *continuous-time* case, see Chapter 6—is to introduce a *new* (monic, ℓ-dependent) polynomial of degree N in the *complex* variable z characterized by the property that its N coefficients evolve in the *discrete time* ℓ as the solutions of the *solvable generation zero system* (7.286). Specifically, we introduce the *generation one polynomial*

$$p_N^{(\mu_1)}(z; \ell) \equiv p_N^{(\mu_1)}\left(z; \vec{y}^{(\mu_1)}(\ell); \tilde{x}^{(\mu_1)}(\ell)\right) = z^N + \sum_{m=1}^{N}\left[y_m^{(\mu_1)}(\ell)\, z^{N-m}\right]$$

$$= \prod_{n=1}^{N}\left[z - x_n^{(\mu_1)}(\ell)\right], \qquad (7.287)$$

such that its *coefficient* N-vector $\vec{y}^{(\mu_1)}(\ell) = \left(y_1^{(\mu_1)}(\ell), \ldots, y_N^{(\mu_1)}(\ell)\right)$ (see Notation 7.1) coincides with an appropriate ordering (prescribed by the index μ_1) of the *unordered* solution set $\tilde{x}(\ell)$ of the *generation zero dynamical system* (7.286) (with $p = 1$). Here the index μ_1 labels the possible orderings of the N *complex* numbers $x_1(\ell), x_2(\ell), \ldots, x_N(\ell)$, implying the definition of the N-vectors

$$\vec{x}_{[\mu_1]}(\ell) \equiv \left(x_{[\mu_1],1}(\ell),\, x_{[\mu_1],2}(\ell), \ldots,\, x_{[\mu_1],N}(\ell)\right). \qquad (7.288a)$$

This index μ_1 takes the $N!$ integer values from 1 to $N!$ (see Notation 7.1), since there are $N!$ different permutations of N different objects, implying the existence of $N!$, generally different, N-vectors $\vec{x}_{[\mu_1]}(\ell)$. And the prescription for $\vec{y}^{(\mu_1)}(\ell)$ mentioned just above, characterizing the *generation one polynomial* (7.287), reads as follows:

$$\vec{y}^{(\mu_1)}(\ell) = \vec{x}_{[\mu_1]}(\ell) \,, \quad y_m^{(\mu_1)}(\ell) = x_{[\mu_1],m}(\ell) \,. \tag{7.288b}$$

Remark 7.1.3. Let us reiterate that throughout our discussion we focus for simplicity on the *generic* case of monic polynomials of degree N in the *complex* variable z the N coefficients of which and likewise the N zeros of which are *all different among themselves*, and correspondingly on dynamical systems describing the evolution in the *discrete time* ℓ of N points moving in the *complex* plane the positions of which at the same time ℓ never coincide. It is indeed evident that this is the *generic* situation, and it will be clear from the following treatment that the occurrence of "particle collisions"—the coincidence of two particle positions at the same time—is an event that can only occur for a set of initial data having *vanishing measure* in the space of such data. ∎

Remark 7.1.4. To allay any possible uneasiness about the notion that a specific permutation labeled by an index μ in the range $1 \leq \mu \leq N!$ identifies a specific order of the N elements x_n of an *unordered* set $\tilde{x}$ of N *complex* numbers x_n, let us provide once more the example of a possible procedure to do so. One begins by *defining* a *specific* ordering assignment of the *unordered* set $\tilde{x}$ of N *different complex* numbers x_n—for instance that characterized by the ("increasing") *lexicographic rule* stipulating that, of two *complex* numbers with *different real* parts, the one with *algebraically smaller real* part comes *first*, and of two *complex* numbers with *equal real* parts, the one with *algebraically smaller imaginary* part comes *first*. After this ordering of the *unordered* set $\tilde{x}$ is thus established—defining the N-vector $\vec{x}_{[1]}$ with components $x_{[1],n}$—one can subsequently apply $N!$ standard sequential reorderings—labeled by the index μ ranging from 2 to $N!$—consisting in $N!$ permutations of the N components $x_{[1],n}$ of the N-vector $\vec{x}_{[1]}$, these permutations being themselves labeled by the value of the index μ according to some standard rule—for instance the standard *lexicographic* ordering of the $N!$ permutations of N different objects (for $N = 3$: $abc, acb, bac, bca, cab, cba$). Note that—see Remark 7.1.3—we always assume to deal with the *generic* case of (monic) polynomials of degree N featuring N *different zeros*, therefore defining new polynomials with N *different coefficients* as detailed above, see (7.288b). ∎

The *generation one dynamical systems* characterize the evolution in the discrete-time ℓ of the *unordered* set $\tilde{x}^{(\mu_1)}(\ell) \equiv \left\{ x_1^{(\mu_1)}(\ell), x_2^{(\mu_1)}(\ell), \ldots, x_N^{(\mu_1)}(\ell) \right\}$ of the N zeros of the polynomial $p_N^{(\mu_1)}(z;\ell)$, see (7.287) with (7.288b). The key formulas (7.282) (with $p = 1$) and (7.288b) show that the equations of motion of these *new* dynamical systems read as follows:

$$\prod_{j=1}^{N} \left[x_n^{(\mu_1)} (\ell + 1) - x_j^{(\mu_1)}(\ell) \right]$$

$$+ \sum_{m=1}^{N} \left\{ \left[x_{[\mu_1],m} (\ell + 1) - x_{[\mu_1],m}(\ell) \right] \left[x_n^{(\mu_1)} (\ell + 1) \right]^{N-m} \right\} = 0 , \quad (7.289a)$$

i.e. they imply that the N (updated) elements $x_n^{(\mu_1)} (\ell + 1)$ of the *unordered* set $\tilde{x}^{(\mu_1)} (\ell + 1) \equiv \left\{ x_1^{(\mu_1)} (\ell + 1), x_2^{(\mu_1)} (\ell + 1), \ldots, x_N^{(\mu_1)} (\ell + 1) \right\}$ are the N zeros of the following polynomial equation of degree N in the *complex* variable z:

$$\prod_{j=1}^{N} \left[z - x_j^{(\mu_1)}(\ell) \right] + \sum_{m=1}^{N} \left\{ \left[x_{[\mu_1],m} (\ell + 1) - x_{[\mu_1],m}(\ell) \right] z^{N-m} \right\} = 0 . \quad (7.289b)$$

Here the coordinates $x_{[\mu_1],m}(\ell)$ appearing in these equations (7.289) are the solutions of the *generation zero dynamical system*, see (7.286), ordered according to the prescription characterized by the value of the index μ_1, as explained above (and see also below). Thus, the quantities $x_{[\mu_1],m} (\ell + 1)$ are the N zeros—ordered according to the prescription identified by the value of the index μ_1—of the polynomial

$$\prod_{j=1}^{N} \left[z - x_j(\ell) \right] + \sum_{m=1}^{N} \left\{ \left[f_m (\vec{y}(\ell); \ell) - y_m(\ell) \right] z^{N-m} \right\} = 0 \quad (7.289c)$$

(see (7.286b) with $p = 1$) where the N quantities $y_m(\ell)$ must be replaced by their expressions (7.281b) in terms of the N coordinates $x_n(\ell)$ (the order of the N coordinates $x_n(\ell)$ is in this context irrelevant, due to the symmetry of the right-hand side of the formulas (7.281b)).

To interpret (7.289) as the *evolution law* for the *unordered* set $\tilde{x}^{(\mu_1)}(\ell)$, we must recall that, via (7.288b), there holds the relation $x_{[\mu_1],m}(\ell) = y_m^{(\mu_1)}(\ell)$, hence, by using the formulas (7.281b), we must make the following replacement in (7.289):

$$x_{[\mu_1],m}(\ell) = y_m^{(\mu_1)}(\ell) = (-1)^m \sum_{1 \leq n_1 \leq n_2 \leq \cdots \leq n_m} \left[x_{n_1}^{(\mu_1)}(\ell) \, x_{n_2}^{(\mu_1)}(\ell) \cdots x_{n_m}^{(\mu_1)}(\ell) \right] .$$

$$(7.289d)$$

The new dynamical system (7.289) is *solvable*, since its solution is provided by the N zeros of the *generation one polynomial* $p_N^{(\mu_1)} (z; \ell)$ given by (7.287), which implies that they can be obtained via an *algebraic* operation. This *generation one* polynomial is itself obtainable by *algebraic* operations since its *coefficients* $y_m^{(\mu_1)} = x_{[\mu_1],m}(\ell)$ are given by the permutation characterized by the index μ_1 of the *unordered* set $\tilde{x}(\ell)$, which is itself obtained as solution of the *solvable* dynamical system (7.286) (with $p = 1$).

Because the evolution prescribed by the new dynamical system (7.289) depends on the assignment of the ordering of the *zeros* of the *generation zero polynomial* (7.281a), see (7.288b), we discuss several possibilities of making this assignment.

One possibility is to make a specific assignment for the ordering prescription, once and for all, corresponding to a specific assignment of the value of the index μ_1: for instance one might assume that all unordered sets $\tilde{x}(\ell)$ be replaced by N vectors $\vec{x}(\ell)$ the N components of which are ordered, say, *lexicographically*. This has the following consequences: (i) consideration is then limited to only a specific one out of many possible *generation one dynamical systems*; (ii) one is then dealing with a dynamical system describing the evolution of *distinguishable* particles, the identity of which is identified by their relative positions in the *complex x*-plane; (iii) the initial data for this specific dynamical system cannot be freely assigned: indeed, their values must be assigned not only consistently with their identities (which is always possible by adjusting the identities to the assigned values), they must moreover be the N zeros of a polynomial of degree N in its *complex* variable z the N *coefficients* of which satisfy themselves the assigned prescription, and this entails a limitation on the corresponding N *zeros*, hence on these *initial values*. Note that an equivalent way to describe this possibility is to let the *initial data* be assigned *arbitrarily* and then to adjust the ordering assignment of the *unordered* sets of data consistently with this initial assignment—but then the very dynamics associated with this point of view would be dependent on the assignment of the *initial data*, which is not consistent with what is usually meant by the definition of a dynamical system as a set of rules—not themselves dependent on the initial data—which determine how the initial data evolve over time ...

Another possibility might be to assign an ℓ-dependent ordering prescription based on *contiguity over time*: given two sets of unordered data $\tilde{x}(\ell)$ and $\tilde{x}(\ell+1)$ one might require that they be ordered so that the distance in the *complex x*-plane between these coordinates at time ℓ and $\ell+1$ having the *same* label be *less* than the distance between the coordinates having different labels, $|x_n(\ell+1) - x_n(\ell)| < |x_n(\ell+1) - x_m(\ell)|$ if $n \neq m$. Clearly this prescription defines *unambiguously* the label assignments (i.e., the particle identities) at time $\ell+1$ corresponding to any label assignment at time ℓ, and viceversa, for any *generic* configuration of $2N$ points $x_n(\ell)$ and $x_n(\ell+1)$ in the *complex x*-plane (i.e., for any *arbitrary* configuration excluding a set of configurations having *zero measure* in the set of all possible configurations); but this is a reasonable prescription only if $|x_n(\ell+1) - x_n(\ell)| << |x_n(\ell+1) - x_m(\ell)|$ if $n \neq m$, namely when—with this assignment—the positions of every particle at time ℓ and $\ell+1$ are much closer to each other than the positions of any two different particles among themselves at time ℓ and at time $\ell+1$ (see examples below).

A third interesting possibility—not discussed any further here—is to assign in a *random* manner—at every step of the *discrete-time* evolution—the prescription to go from the *unordered* set of the N *zeros* of the *generation zero* polynomial to the *ordered* set of the N *coefficients* of the *generation one* polynomial, thereby producing a dynamical system featuring a *random* evolution.

Up to now we have discussed only the *first* iteration of our approach—and this justified our notation μ_1 rather than just μ for the relevant parameter identifying the prescription characterizing the transition (7.288) from the N *zeros* of the *generation zero polynomial* to the N *coefficients* of the *generation one polynomial*, yielding the identification of *solvable "generation one" dynamical systems*. Further iterations could be performed, yielding *new solvable discrete-time dynamical systems*: these developments—which are not detailed here—are rather obvious, given the analogies with the treatment provided in the *continuous-time* context, see Chapter 6, and the new features of the *discrete-time* context discussed above.

We complete this Section 7.1 with the following important remark:

> **Remark 7.1.5.** A *solvable* dynamical system generally inherits some properties from the *seed solvable* dynamical system generating it. For instance, if the *seed solvable* dynamical system is *isochronous* respectively *asymptotically isochronous* with period L, then the *generation k solvable* systems—characterizing the evolution of *unordered* sets of N indistinguishable points—are as well *isochronous* respectively *asymptotically isochronous* with period L or an *integer multiple* of L (at most $(\nu_{\text{Max}}(N))^k L$: see Section 2.3), for all $k = 1, 2, \ldots$ To elaborate, suppose that the solution $\vec{y}(\ell)$ of the *seed system* has the *isochronicity* property $\vec{y}(\ell + L) = \vec{y}(\ell)$, where L is a *fixed positive integer*. Then the unordered solution $\tilde{x}(\ell)$ of the *generation zero* system has the same property: $\tilde{x}(\ell + L) = \tilde{x}(\ell)$. The solution $\tilde{x}^{(\mu_1)}(\ell)$ of the *generation one* system is also *isochronous* with period L_1: $\tilde{x}^{(\mu_1)}(\ell + L_1) = \tilde{x}^{(\mu_1)}(\ell)$, where $L_1 = L$ if the *lexicographic* rule (or one of its $N!$ variants) was used to order the zeros $\tilde{x}(\ell)$ of the *generation zero* system—the same rule for all values of ℓ— and $L_1 = pL$ with p a *positive integer* ($p \le \nu_{\text{Max}}(N)$) if the *contiguity* rule was instead applied. Similarly, if the solution $\vec{y}(\ell)$ of the *seed system* has the *asymptotic isochronicity* property $\vec{y}(\ell + L) - \vec{y}(\ell) \to 0$ as $\ell \to \infty$, then the solution $\tilde{x}(\ell)$ of the *generation zero* system has the same property: $\tilde{x}(\ell + L) - \tilde{x}(\ell) \to 0$ as $\ell \to \infty$, while the solution $\tilde{x}^{(\mu_1)}(\ell)$ of the *generation one* system is *asymptotically isochronous* with a period L_1 that might be an *integer multiple* of L, depending on the ordering rules used at every step of the discrete-time evolution. ∎

The validity of this Remark 7.1.5 is justified by considerations sufficiently analogous to those made in the *continuous-time* context, see Chapter 6, not to warrant their repetition here. These considerations are also illustrated by the examples reported in the following Section 7.2, for instance see below Remark 7.2.1.

Let us end this Section 7.1 by proving the key formula (7.282). This is in fact an immediate consequence of definition (7.281a) of the polynomial $p_N(z; \ell)$. Indeed, let p be a *positive integer*. The first of the two equalities (7.281a) then implies

$$p_N(z; \ell + p) - p_N(z; \ell) = \sum_{m=1}^{N} \left\{ \left[y_m(\ell + p) - y_m(\ell) \right] z^{N-m} \right\}, \qquad (7.290a)$$

and, via the second of the two equalities (7.281a), this formula reads

$$\prod_{j=1}^{N} \left[z - x_j(\ell + p) \right] - \prod_{j=1}^{N} \left[z - x_j(\ell) \right] = \sum_{m=1}^{N} \left\{ \left[y_m(\ell + p) - y_m(\ell) \right] z^{N-m} \right\}.$$

$$(7.290b)$$

By setting in this formula $z = x_n(\ell + p)$ one obtains (7.282). Q. E. D.

Remark 7.1.6. Clearly, by setting $z = x_n(\ell)$ in (7.290b), one obtains the alternative formula

$$\prod_{j=1}^{N} \left[x_n(\ell) - x_j(\ell + p) \right] = \sum_{m=1}^{N} \left\{ \left[y_m(\ell + p) - y_m(\ell) \right] \left[x_n(\ell) \right]^{N-m} \right\}. \qquad (7.291)$$

7.2 Examples

In Section 7.2 we illustrate the findings reported above (in this Chapter 7) via some examples.

Notation 7.2.1. In this Section 7.2, we often omit to indicate explicitly the ℓ-dependence of the quantities under consideration, and we use a superimposed "hat" to denote a unit updating of the discrete-time variable ℓ, hence, for instance, $\hat{y}_m \equiv y_m(\ell + 1)$. ∎

Example 7.2.1. As *seed dynamical system* take the simple first-order dynamical system

$$\hat{y}_m = a_m\, y_m + b_m\,, \qquad (7.292a)$$

corresponding to (7.284) with $p = 1$ and

$$f_m(\vec{y}) = a_m\, y_m + b_m\,. \qquad (7.292b)$$

Note that this dynamical system is a simpler variant of (7.285a) because the $2N$ parameters a_m and b_m are assumed to be ℓ-independent.

It is easily seen that the solution of these equations of motion reads

$$y_m(\ell) = (a_m)^\ell\, y_m(0) + \left[\frac{(a_m)^\ell - 1}{a_m - 1} \right] b_m\,. \qquad (7.293)$$

Hereafter, for simplicity, we assume that $a_m \neq 1$ for all values of m in the range from 1 to N.

The *generation zero* system constructed from the *seed system* (7.292a) via the method described in Section 7.1, see (7.282), reads

$$\prod_{j=1}^{N} \left[x_n \left(\ell + 1 \right) - x_j(\ell) \right]$$

$$+ \sum_{m=1}^{N} \left\{ \left[(a_m - 1)y_m(\ell) + b_m \right] \left[x_n \left(\ell + p \right) \right]^{N-m} \right\} = 0, \qquad (7.294)$$

where $y_m(\ell)$ are expressed in terms of $\tilde{x}(\ell)$ via (7.281b). This system describes the evolution of the *unordered* set $\tilde{x}(\ell)$ of the zeros of the monic polynomial with the coefficients $y_m(\ell)$ given by (7.293), see the discussion below for the case where $N = 2$. ∎

Remark 7.2.1. Clearly if

$$a_m = \exp \left(\frac{2 \, \pi \, \mathbf{i} \, q_m}{k_m} \right) \qquad (7.295)$$

with k_m an arbitrary *positive integer*, q_m an *arbitrary integer* (*coprime* to k_m), then $y_m(\ell)$ (see (7.293)) is *isochronous* with period $L_m = k_m$, namely for any arbitrary initial datum $y_m(0)$ there holds the periodicity property $y_m \left(\ell + L_m \right) = y_m(\ell)$. Note that (7.295) implies $|a_m| = 1$. While if $|a_m| < 1$, then—again, for any arbitrary initial datum $y_m(0)$—there obtains from (7.293) the asymptotic limit $y_m (\infty) = b_m/ \left(1 - a_m \right)$.

Clearly the *generic* solutions of the *generation zero model* (7.294) obtained from this *seed model* (7.292a) via the technique described in Section 7.1 have the following remarkable properties: (i) they are *isochronous* with period $L = K$, where K is the *Minimum Common Multiple* of the *positive integers* $k_1, \ldots, k_N$, provided that all the N parameters a_m satisfy condition (7.295); (ii) they are *asymptotically isochronous* with asymptotic period $L = K$ if some (at least one but not all) of the parameters a_m—say, the parameters $a_{\tilde{s}}$—have modulus less than *unity*, $|a_{\tilde{s}}| < 1$, and the remaining parameters $a_{\hat{s}}$ satisfy condition (7.295); in this case, K is the the *Minimum Common Multiple* of the corresponding indices $k_{\hat{s}}$; (iii) they converge *asymptotically* to fixed values (independent of the initial data) if $|a_m| < 1$ for all $m = 1, \ldots, N$; (iv) if some among the parameters a_m, say a_j, have modulus larger than unity, $|a_j| > 1$, for *generic* initial data some of the N coordinates $x_n(\ell)$ *diverge asymptotically* as $\ell \to \infty$ and others converge to a *finite value*, as implied by the findings reported in Appendix G ("Asymptotic behavior of the zeros of a polynomial whose coefficients diverge exponentially") of [20]. ∎

Let us then discuss the *generation zero* dynamical system that obtains from the *seed system* (7.292a), for the case where $N = 2$. To construct this *generation zero* system, consider the *generation zero* polynomial given by

$$p_2(z) = z^2 + y_1(\ell)\, z + y_2(\ell) = [z - x_1(\ell)]\ [z - x_2(\ell)]\,, \tag{7.296}$$

where $y_1(\ell)$, $y_2(\ell)$ are given by (7.292a) with $N = 2$. Note that the *ordered pair* of the coefficients $\vec{y}(\ell) = (y_1(\ell), y_2(\ell))$ determines the *unordered set* of the zeros $\tilde{x}(\ell) \equiv \{x_1(\ell),\, x_2(\ell)\}$. [Note that here and hereafter the notation $\{u,\, v\}$ indicates the *unordered* set of the two numbers u and v]. The evolution of the latter *unordered* set is given by the dynamical system

$$\tilde{x}(\ell + 1) = \{g_1\,(\tilde{x}(\ell))\,,\ g_2\,(\tilde{x}(\ell))\}\,, \tag{7.297a}$$

where (for $m = 1, 2$)

$$g_m(\tilde{x}) \equiv g_m\,(\{x_1, x_2\}) = \frac{1}{2}\,[a_1\,(x_1 + x_2) - b_1]$$
$$+ \frac{(-1)^m}{2}\,\{[a_1\,(x_1 + x_2) - b_1]^2 - 4\,b_2 - 4\,a_2\,x_1\,x_2\}^{1/2}\,. \tag{7.297b}$$

This system (7.297) is the particular case of system (7.294) for $N = 2$. Its solution with the initial condition $\tilde{x}(0) = \{x_1(0),\, x_2(0)\}$ reads

$$\tilde{x}(\ell) = \{x_1(\ell),\, x_2(\ell)\}\,, \tag{7.298a}$$

where

$$x_m(\ell) = \frac{1}{2}\left[-y_1(\ell) + (-1)^m\,\{[y_1(\ell)]^2 - 4\,y_2(\ell)\}^{1/2}\right]\,, \tag{7.298b}$$

with $y_m(\ell)$ given by (7.293) with $y_1(0) = -x_1(0) - x_2(0)$ and $y_2(0) = x_1(0)x_2(0)$. In more explicit form,

$$x_m(\ell) = \frac{1}{2}\left\{(a_1)^\ell\,[x_1(0) + x_2(0)] - \left(\frac{(a_1)^\ell - 1}{a_1 - 1}\right)b_1\right\}$$
$$+ (-1)^m\,\frac{1}{2}\left\{\left[(a_1)^\ell\,[x_1(0) + x_2(0)] - \left(\frac{(a_1)^\ell - 1}{a_1 - 1}\right)b_1\right]^2\right.$$
$$\left. - 4\left[(a_2)^\ell\,x_1(0)\,x_2(0) + \left(\frac{(a_2)^\ell - 1}{a_2 - 1}\right)b_2\right]\right\}^{1/2}\,,$$
$$\ell = 1, 2, 3, \ldots. \tag{7.298c}$$

Remark 7.2.2. If the generic complex number z is defined via its modulus and phase as follows, $z = |z|\,\exp(\mathbf{i}\,\varphi)$ with $0 < \varphi \le 2\pi$, its square-root $\sqrt{z}$ is defined up to a sign ambiguity, say $\sqrt{z} = \pm\sqrt{|z|}\,\exp(\mathbf{i}\,\varphi/2)$, with $\sqrt{|z|} > 0$. But, above and hereafter, we assume for definiteness—irrelevant as this is—that the notation $z^{1/2}$ indicates a specific determination of the square-root of z, say that with *positive real part*, and, *if the real part vanishes, with positive imaginary part*. ∎

Example 7.2.2. Let us now report the *generation one* dynamical systems that obtain from the *seed system* (7.292a), again taking $N = 2$ for simplicity. Recall that for $N = 2$ the *generation one* polynomials are given by

$$p_2^{(\mu_1)}(z) = z^2 + y_1^{(\mu_1)}(\ell)\, z + y_2^{(\mu_1)}(\ell) = \left[z - x_1^{(\mu_1)}(\ell)\right]\left[z - x_2^{(\mu_1)}(\ell)\right], \quad \mu_1 = 1, 2,$$
(7.299)

see (7.287), where the ordered pair $\vec{y}^{(\mu_1)}(\ell) = \left(y_1^{(\mu_1)}(\ell), y_2^{(\mu_1)}(\ell)\right)$ equals an appropriately ordered pair $x_1(\ell), x_2(\ell)$. For example, we may choose the *lexicographic* order of $x_1(\ell)$, $x_2(\ell)$ if $\mu_1 = 1$, and the other *(anti-lexicographic)* order if $\mu_1 = 2$.

The dynamical system for the *unordered* pair $\tilde{x}^{(\mu_1)}(\ell) = \left\{x_1^{(\mu_1)}(\ell),\, x_2^{(\mu_1)}(\ell)\right\}$ is then given by

$$\tilde{x}^{(\mu_1)}(\ell + 1) = \left\{g_1^{(\mu_1)}\left(\tilde{x}^{(\mu_1)}(\ell)\right),\, g_2^{(\mu_1)}\left(\tilde{x}^{(\mu_1)}(\ell)\right)\right\},$$
(7.300a)

where

$$g_m^{(\mu_1)}(\tilde{x}) \equiv g_m(\{x_1, x_2\}) = -\frac{1}{2}\gamma_-^{(\mu_1)}(\tilde{x})$$
$$+ \frac{1}{2}(-1)^m \left\{\left[\gamma_-^{(\mu_1)}(\tilde{x})\right]^2 - 4\,\gamma_+^{(\mu_1)}(\tilde{x})\right\}^{1/2}$$
$$m = 1, 2, \quad \mu_1 = 1, 2,$$
(7.300b)

with the pair $\left(\gamma_-^{(\mu_1)}(\tilde{x}),\, \gamma_+^{(\mu_1)}(\tilde{x})\right)$ being equal to the ordering of the pair $\{\alpha(\tilde{x}) - (-1)^{\mu_1}\beta(\tilde{x}),\, \alpha(\tilde{x}) + (-1)^{\mu_1}\beta(\tilde{x})\}$ that corresponds to the value of the index μ_1, where

$$\alpha(\tilde{x}) = \frac{1}{2}\left[a_1(-x_1 - x_2 + x_1 x_2) - b_1\right],$$
(7.301a)

$$\beta(\tilde{x}) = \left\{\left[\alpha(\tilde{x})\right]^2 - b_2 + a_2 x_1 x_2(x_1 + x_2)\right\}^{1/2}.$$
(7.302)

The solution of this dynamical system (7.300) with the initial condition $\tilde{x}(0) = \{x_1(0),\, x_2(0)\}$ is given by

$$\tilde{x}^{(\mu_1)}(\ell) = \left\{x_1^{(\mu_1)}(\ell),\, x_2^{(\mu_1)}(\ell)\right\},$$
(7.303a)

where

$$x_m^{(\mu_1)}(\ell) = \frac{1}{2}\left[-y_1^{(\mu_1)}(\ell) + (-1)^m\left\{\left[y_1^{(\mu_1)}(\ell)\right]^2 - 4\,y_2^{(\mu_1)}(\ell)\right\}^{1/2}\right]$$
(7.303b)

and the numbers $y_m^{(\mu_1)}(\ell)$ are given by the formulas (7.298c) for $x_m(\ell)$, with $x_m(0)$ replaced by $y_m^{(\mu_1)}(0)$, where

$$y_1^{(\mu_1)}(0) = -x_1(0) - x_2(0), \quad y_2^{(\mu_1)}(0) = x_1(0) x_2(0).$$
(7.303c)

As before, $\mu_1 = 1$ indicates the *lexicographic* order of the pair $y_1^{(\mu_1)}(\ell)$, $y_2^{(\mu_1)}(\ell)$, while $\mu_1 = 2$ indicates the other (*anti-lexicographic*) order.

In summary, given the initial condition $\{x_1(0),\ x_2(0)\}$, we can solve the dynamical system (7.300) as follows. *First*, we order the initial condition to ensure that the pair $(x_1(0),\ x_2(0))$ is in *lexicographic* order. *Second*, we compute $y_m^{(\mu_1)}(0)$ by formulas (7.303c) and assign $\mu_1 = 1$ if the pair $(-x_1(0) - x_2(0),\ x_1(0)\ x_2(0))$ turns out to be in *lexicographic* order and $\mu_1 = 2$ otherwise. *Third*, we compute $y_m^{(\mu_1)}(\ell)$ using the formulas (7.298c) for $x_m(\ell)$, with $x_m(0)$ replaced by $y_m^{(\mu_1)}(0)$, while ensuring that each pair $\left(y_1^{(\mu_1)}(\ell),\ y_2^{(\mu_1)}(\ell)\right)$ is ordered according to the value of μ_1 chosen in the previous step. *Fourth*, we compute $x^{(\mu_1)}(\ell)$ using formulas (7.303). ∎

Example 7.2.3. In this example we take the following *solvable second-order discrete-time* dynamical system as *seed* system:

$$y_m(\ell + 2) = a_m(\ell)\,\frac{\left[y_m(\ell + 1)\right]^2}{y_m(\ell)} + b_m(\ell)\,y_m(\ell + 1)\,, \qquad (7.304)$$

where $a_m(\ell)$ and $b_m(\ell)$ are some functions of ℓ. Via the substitution

$$u_m(\ell) = \frac{y_m(\ell + 1)}{y_m(\ell)}\,, \qquad (7.305)$$

we find that the solution of system (7.304) with the initial conditions $y_m(0),\ y_m(1)$ is given (see Notation 7.1) by

$$y_m(\ell) = y_m(0)\,\prod_{j=0}^{\ell-1} u_m(j)\,, \qquad (7.306a)$$

where

$$u_m(\ell) = \frac{y_m(1)}{y_m(0)}\,\prod_{j=0}^{\ell-1}\left[a_m(j)\right] + \sum_{k=0}^{\ell-1}\left\{\left[\prod_{j=k+1}^{\ell-1} a_m(j)\right] b_m(k)\right\}\,. \qquad (7.306b)$$

Hence the *discrete-time* dynamical system characterized by the *second-order* equations of motion

$$\prod_{j=1}^{N}\left[x_n(\ell + 2) - x_j(\ell)\right]$$

$$+ \sum_{m=1}^{N}\left\{\left[a_m(\ell)\,\frac{\left[y_m(\ell + 1)\right]^2}{y_m(\ell)} + b_m(\ell)\,y_m(\ell + 1) - y_m(\ell)\right]\,[x_n(\ell + 2)]^{N-m}\right\}$$

$$= 0\,, \qquad (7.307)$$

with $y_m(\ell)$ and $y_m(\ell + 1)$ in the right-hand side replaced with their appropriate expressions depending on $\tilde{x}(\ell)$ and $\tilde{x}(\ell + 1)$, see (2.4), is *solvable*: indeed its solutions are provided by the N zeros of the polynomial $z^N + \sum_{m=1}^{N}\left[y_m(\ell)\,z^{N-m}\right]$ the N coefficients $y_m(\ell)$ of which are given by the formulas (7.306).

For $N = 2$ these discrete-time equations of motion read as follows:

$$[x_n(\ell+2)]^2 - \left\{ \frac{a_1(\ell)\ [x_1(\ell+1) + x_2(\ell+1)]}{x_1(\ell) + x_2(\ell)} + b_1(\ell) \right\} \cdot$$
$$\cdot\ [x_1(\ell+1) + x_2(\ell+1)]\ x_n(\ell+2)$$
$$+ \left[\frac{a_2(\ell)\ [x_1(\ell+1)\ x_2(\ell+1)]}{x_1(\ell)\ x_2(\ell)} + b_2(\ell) \right] x_1(\ell+1)\ x_2(\ell+1)$$
$$+ x_1(\ell)\ x_2(\ell) = 0\,, \quad n = 1,2\,. \tag{7.308}$$

7.3 Outlook

In this last section of Chapter 7 we firstly reiterate a remark—analogous to the previous Remark 6.1.1—relevant to the entire content of this book and we then tersely outline a variant of the *discrete-time* approach.

> **Remark 7.3.1.** A simple way to "generalize" any dynamical system is to perform a change of dependent and independent variables; but generally the "new" models obtained in such a way from a known model are not considered *really new*. It might therefore be inferred that the technique described in this chapter—indeed in this book—are not really yielding *new* solvable dynamical systems, since the main tool employed—the relation between the *coefficients* and the *zeros* of a polynomial—may well be considered just a change of dependent variables for the dynamical systems under consideration. But this criticism conflicts with the observation that essentially *all solvable* dynamical system can be reduced, by *appropriate* changes of variables, to *trivial* evolutions. The rub is in the identification of the *appropriate* changes of variables. Hence the emergence of the "inverse" approach: to start from certain changes of variables—in particular, those relating the (time-dependent) *coefficients* and the (time-dependent) *zeros* of (time-dependent, monic) polynomials—and to then try and identify the dynamical systems *solvable* via this kind of transformation of dependent variables. To those who consider such an "inverse" approach a kind of cheating, we can only reply by begging them to ponder what is written in the Foreword (see, in particular, page VII) of the book [20] to justify this approach (indeed, amply practiced both in that book and in most other publications on *solvable/integrable* dynamical systems, including the present one). ■

Let us finally indicate a possible future development, which we leave as a task for the interested reader. An analogous approach to that discussed in this paper—but leading to *solvable* dynamical systems in "q-discrete time," that is, being characterized by "q-difference" equations of motion rather than by "difference" equations of motion—obtains by taking as point of departure, instead of the polynomial formula (7.281a), the following definition:

$$p_N(z;\ q;\ \ell) = z^N + \sum_{m=1}^{N}\left[y_m\left(q^\ell\right)\ z^{N-m}\right] = \prod_{n=1}^{N}\left[z - x_n\left(q^\ell\right)\right], \tag{7.309}$$

with q an *arbitrarily assigned* parameter ($q \neq 1$). It is then easily seen—again, by a quite analogous treatment to that provided in Section 7.1—that the key formula (7.282) gets replaced by the relation

$$\prod_{j=1}^{N}\left[x_n\left(q^{\ell+p}\right) - x_j\left(q^\ell\right)\right] + \sum_{m=1}^{N}\left\{\left[y_m\left(q^{\ell+p}\right) - y_m\left(q^\ell\right)\right]\ \left[x_n\left(q^{\ell+p}\right)\right]^{N-m}\right\} = 0,$$
$$\tag{7.310}$$

where p is an *arbitrary integer*. And a simple example of *first-order seed* system to generate other *nonlinear solvable* dynamical systems in "q-discrete time" reads as follows:

$$y_m\left(q^{\ell+1}\right) = a_m\ y_m\left(q^\ell\right) + b_m, \tag{7.311a}$$

since the solution of this evolution equation is explicitly given in terms of the initial values $y_m(1)$ (corresponding to $\ell = 0$) by the following formula:

$$y_m\left(q^\ell\right) = (a_m)^\ell\ y_m(1) + \left[\frac{(a_m)^\ell - 1}{a_m - 1}\right]\ b_m. \tag{7.311b}$$

7.N Notes on Chapter 7

For the results reported in Chapter 7, see [6], where additional discussions, and also plots, of some solutions of the examples reported in this Chapter 7 are provided, as well as references to previous works on *discrete-time* dynamical systems.

An analogous remark to that made in the next to last paragraph of Section 3.N can be made in connection with *discrete-time* evolutions: see F. Calogero, "Solvable nonlinear discrete-time evolutions and Diophantine findings," J. Nonlinear Math. Phys. **25**, 1–3 (2018).

8

Outlook

In this terse Chapter 8 we indicate some further developments—besides the few already mentioned above—that are naturally suggested as follow-up to the results reported in this book.

(i) Techniques have been provided above to identify in the context of *classical* mechanics many N-body problems which are amenable to *exact* treatment, including many *Hamiltonian* models and many that feature *isochronous* evolutions. An interesting idea—in these *Hamiltonian* cases, and particularly in those featuring *isochronous* evolutions—is to investigate the corresponding models in the context of *quantum* mechanics. Also in view of the natural conjecture that those *Hamiltonian* systems that, in the context of *classical* mechanics, yield *isochronous* time evolutions, do then feature, in the context of *quantum* mechanics, *equispaced* energy spectra.

(ii) The *solvable N-body problems* identified by the techniques discussed above generally feature solutions that, for $N \leq 4$, can be obtained in *completely explicit* form (although for $N > 2$ the relevant formulas are not particularly transparent). But what about the behavior of these *solvable N-body models* for *very large* values of N, or even in the $N \to \infty$ limit using the formalism of (*classical* or *quantal*) *statistical* mechanics? The interest of this development is underlined by the possibility to obtain in this manner interesting mathematical findings, as demonstrated in an analogous case by Gallavotti and Marchioro [73]. A related interesting development is the investigation of the relationship among the *infinitely many coefficients* $y_m(t)$ and *zeros* $x_n(t)$ of time-dependent *entire* functions $F(z;t)$ of the *complex* variable z admitting the following, absolutely convergent, representations both as an *infinite* sum and as an *infinite* product:

$$F(z;t) = 1 + \sum_{m=1}^{\infty} \left[y_m(t) \ z^m \right] = \prod_{n=1}^{\infty} \left[1 - \frac{z}{x_n(t)} \right] ; \qquad (8.312)$$

for recent progress in this direction see [45].

(iii) The basic tool of the developments reported in this book has been the relationship among the time-evolution of the *N coefficients* $y_m(t)$ and the *N zeros* $x_n(t)$ of a time-dependent (monic) polynomial of degree *N* in its argument *z*: see (2.3). This typically allowed to identify time-evolution models—be they *solvable* systems of *nonlinear* ODEs or PDEs—with equations of motion featuring elementary functions and variables embedded in the *complex* plane (or equivalently in the 2-dimensional Cartesian plane). *Extensions* of this approach using as basic tools *more general* functions than polynomials—such as *trigonometric* or *elliptic* functions—and to a *multidimensional* environment are interesting possibilities. The treatment of special classes of polynomials, such as those featuring multiple zeros or those conveniently expressed as superpositions of classical *orthogonal* polynomials, are also worthy of investigation: for recent progress in this direction see [7], [8].

(iv) A large area of potential *applications* of the *solvable* models described in this book—and of the large universe of *additional solvable* models which can be identified by the techniques described in it—is open: perhaps in fields like population dynamics, ecology, chemical reactions, economics, you name it, mainly by focusing on systems of *first-order* nonlinear ODEs—an aspect that, because of its *mathematical simplicity* and instead of the *physical relevance* of *second-order* nonlinear ODEs (i.e., *Newtonian* equations of motion), has been somewhat *underplayed* in this book.

8.N Notes on Chapter 8

For the *quantization* of *isochronous* models (see above item (i) in this Chapter 8) and the conjecture that they yield *equispaced energy spectra* see [57], [22], [24], [47], [60], [61], [62], [63]. But in this connection the possibility should be kept in mind that, to dynamical systems characterized by Newtonian equations of motion yielding *isochronous* evolutions, Hamiltonian systems might be associated yielding the *same* (*isochronous*) evolution in configuration space (i.e., for the *canonical coordinates*), but a *nonisochronous* evolution for the corresponding *canonical momenta*. Such Hamiltonians do *not* yield *equispaced* spectra after quantization: for a recent discussion of this phenomenology see [68].

Indeed the conjecture that a Hamiltonian yielding *isochronous* evolution in a *classical* context feature an *equispaced* spectrum in the *quantal* context is only tenable if that *isochronous* evolution is featured both by the *canonical variables* and by the corresponding *canonical momenta*: see [68].

For hints on how to pursue the extensions to functions other than polynomials—as mentioned in item (iii) of this Chapter 8—see [50], Sections 2.3.5 and 2.3.6.3 of [20], and [69], [51], [52].

Appendix

Complex Numbers and Real 2-Vectors

In this Appendix we report, for the convenience of the reader, some standard formulas detailing the *equivalence* (denoted by the symbol $\Leftrightarrow$) among *complex* numbers, such as Z, and *real* 2-vectors, such as $\vec{R}$. We trust the notation to be self-explanatory: $\mathbf{i}$ is the *imaginary* unit, $\mathbf{i}^2 = -1$; X and Y are 2 *real* numbers; Z is a *complex* number; $\vec{R}$ is a *real* 2-vector.

$$Z \equiv X + \mathbf{i}Y \quad \Leftrightarrow \quad \vec{R} \equiv (X,\, Y)\,, \tag{A.313}$$

$$\vec{R}_1 \cdot \vec{R}_2 = X_1\, X_2 + Y_1\, Y_2\,, \tag{A.314}$$

$$|Z|^2 = X^2 + Y^2 = R^2 = \left(\vec{R} \cdot \vec{R}\right)\,, \tag{A.315}$$

$$\mathbf{i}Z = -Y + \mathbf{i}\,X \quad \Leftrightarrow \quad \hat{R} \equiv (-Y,\, X)\,, \tag{A.316}$$

$$
\begin{aligned}
Z_1\, (Z_2)^* &= X_1\, X_2 + Y_1\, Y_2 \; + \mathbf{i}\,(X_1\, Y_2 - X_2\, Y_1) \\
&= \left(\vec{R}_1 \cdot \vec{R}_2\right) - \mathbf{i}\left(\vec{R}_1 \cdot \hat{R}_2\right)
\end{aligned} \tag{A.317}
$$

$$\frac{Z_1 Z_2}{Z_3} \quad \Leftrightarrow \quad \frac{\vec{R}_1\left(\vec{R}_2 \cdot \vec{R}_3\right) + \vec{R}_2\left(\vec{R}_1 \cdot \vec{R}_3\right) - \vec{R}_3\left(\vec{R}_1 \cdot \vec{R}_2\right)}{(R_3)^2}\,, \tag{A.318}$$

$$
\begin{aligned}
\frac{(Z_1)^2\, Z_2}{Z_3\, Z_4} \Leftrightarrow\; &(R_3)^{-2}\,(R_4)^{-2}\, \cdot \\
\cdot\, \Big\{ &\left[\left(\vec{R}_1 \cdot \vec{R}_3\right)\left(\vec{R}_1 \cdot \vec{R}_4\right) - \left(\vec{R}_1 \cdot \hat{R}_3\right)\left(\vec{R}_1 \cdot \hat{R}_4\right)\right]\vec{R}_2 \\
- &\left[\left(\vec{R}_1 \cdot \vec{R}_3\right)\left(\vec{R}_1 \cdot \hat{R}_4\right) + \left(\vec{R}_1 \cdot \hat{R}_3\right)\left(\vec{R}_1 \cdot \vec{R}_4\right)\right]\hat{R}_2 \Big\}\,.
\end{aligned} \tag{A.319}
$$

Note that the vector $\hat{R} = (-Y,\, X)$ in the XY-plane is equivalently related to the vector $\vec{R} = (X,\, Y)$ in the XY-plane by the 3-dimensional formula

$$\hat{R} = \check{z} \wedge \vec{R}, \tag{A.320}$$

where, in 3-dimensional space notation, $\hat{R} = (-Y,\, X,\, 0)$, $\vec{R} = (X,\, Y,\, 0)$, and $\check{z} = (0,\, 0,\, 1)$ is the *unit* 3-vector orthogonal to the XY plane, while the symbol $\wedge$ denotes the standard 3-dimensional *vector* product. In other words, the 2-vector $\hat{R}$ is the vector $\vec{R}$ rotated counterclockwise around the origin in the Cartesian plane by a $\pi/2$ angle.

References

[1] J. Arlind, M. Bordemann, J. Hoppe, and C. Lee. "Goldfish geodesics and Hamiltonian reduction of matrix dynamics." *Lett. Math. Phys.* **84**: 89–98. (2008).

[2] O. Bihun and F. Calogero. "Solvable many-body models of goldfish type with one-, two- and three-body forces." SIGMA **9**(059) (18 pages). (2013). http://arxiv.org/abs/1310.2335.

[3] O. Bihun and F. Calogero. "Novel solvable many-body problems." *J. Nonlinear Math. Phys.* **23**: 190–212. (2016). DOI: 10.1080/14029251.2016.1161260.

[4] O. Bihun and F. Calogero. "A new solvable many-body problem of goldfish type." *J. Nonlinear Math. Phys.* **23**: 28–46. (2016); arXiv:13749 [math-ph].

[5] O. Bihun and F. Calogero. "Generations of monic polynomials such that the coefficients of each polynomial of the next generation coincide with the zeros of a polynomial of the current generation, and new solvable many-body problems." *Lett. Math. Phys.* **106**(7), 1011–1031. (2016). DOI: 10.1007/s11005-016-0836-8.

[6] O. Bihun and F. Calogero. "Generations of *solvable discrete-time* dynamical systems," *J. Math. Phys.* **58**: 052701 (2017). DOI: 10.1063/1.4928959.

[7] O. Bihun and F. Calogero. "Time-dependent polynomials with *one double* root, and related new solvable systems of nonlinear evolution equations," *Qual. Theory Dyn. Syst.* (in press) doi.org/10.1007/s12346-018-0282-3; http://arxiv.org/abs/1806.07502.

[8] O. Bihun and F. Calogero. "Zeros of time-dependent superpositions of orthogonal polynomials, and related new solvable systems of nonlinear evolution equations." (in preparation).

[9] O. Bihun and D. Fulghesu. "Polynomials whose coefficients coincide with their zeros." *Aequationes mathematicae* **92**(3), 453–470 (2018). DOI: 10.1007/s000100-018-0546-7. arXiv:1705.02057v1[Math.CA]5May2017.

[10] M. Bruschi and F. Calogero. "Novel solvable variants of the goldfish many-body model." *J. Math. Phys.* **47**: 022703 (2006) (25 pages).

[11] M. Bruschi and F. Calogero. "Goldfishing: a new solvable many-body problem." *J. Math. Phys.* **47**: 042901:1–35. (2006).

[12] M. Bruschi and F. Calogero. "A convenient expression of the time-derivative $z_n^{(k)}(t)$, of arbitrary order k, of the zero $z_n(t)$ of a time-dependent polynomial $p_N(z;t)$ of arbitrary degree N in z, and solvable dynamical systems." *J. Nonlinear Math. Phys.* **23**: 474–485. (2016).

[13] M. Bruschi, F. Calogero, and F. Leyvraz. "A large class of solvable discrete-time many-body problems." *J. Math. Phys.* **55**: 082703 (9 pages) (2014). DOI: http://dx.doi.org/10.1063/1.4891760.

[14] M. Bruschi and F. Calogero. "A convenient expression of the time-derivative $z_n^{(k)}(t)$, of arbitrary order k, of the zero $z_n(t)$ of a time-dependent polynomial $p_N(z;t)$ of arbitrary degree N in z, and solvable dynamical systems." *J. Nonlinear Math. Phys.* **23**: 474–485. (2016).

[15] F. Calogero. "Solution of the one-dimensional N-body problem with quadratic and inversely-quadratic pair potentials." *J. Math. Phys.* **12**: 419–436. Erratum: 1996. Journal of Mathematical Physics 37: *J. Math. Phys.* **37**, 3646. (1996).

[16] F. Calogero. "Motion of poles and zeros of special solutions of nonlinear and linear partial differential equations, and related 'solvable' many body problems." *Nuovo Cimento* **43B**: 177–241. (1978).

[17] F. Calogero. "Why are certain nonlinear PDEs both widely applicable and integrable?" In *What is integrability?*, edited by V. E. Zakharov, 1–62. Heidelberg: Springer. (1991).

[18] F. Calogero. "Integrable nonlinear evolution equations and dynamical systems in multidimensions." in: *KdV '95*, Proceedings of the International Symposium held in Amsterdam, The Netherlands, April 23-26,1995 (edited by M. Hazewinkel, H.W. Capel and E.M. de Jager).

[19] F. Calogero. "A class of integrable Hamiltonian systems whose solutions are (perhaps) all completely periodic." *J. Math. Phys.* **38**: 5711–5719. (1997).

[20] F. Calogero. *Classical many-body problems amenable to exact treatments*, Lecture Notes in Physics Monograph **m66**. Heidelberg: Springer. (2001). (749 pages).

[21] F. Calogero. "The 'neatest' many-body problem amenable to exact treatments (a 'goldfish'?)." *Physica D:* **152–153**, 78–84. (2001).

[22] F. Calogero. "On the quantization of two other nonlinear harmonic oscillators." 2003. *Phys. Lett.* A **319**: 240–245. (2003).

[23] F. Calogero. "Solution of the goldfish N-body problem in the plane with (only) nearest-neighbor coupling constants all equal to minus one half." *J. Nonlinear Math. Phys.* **11**: 102–112. (2004).

[24] F. Calogero. "On the quantization of yet another two nonlinear harmonic oscillators." *J. Nonlinear Math. Phys.* **11**: 1–6. (2004).

[25] F. Calogero. "A technique to identify solvable dynamical systems, and a solvable generalization of the goldfish many-body problem." *J. Math. Phys.* **45**: 2266–2279. (2004).

[26] F. Calogero. "A technique to identify solvable dynamical systems, and another solvable extension of the goldfish many-body problem." *J. Math. Phys.* **45**: 4661–4678. (2004).

[27] F. Calogero. *Isochronous systems*. Oxford University Press: Oxford. (2008) (264 pages) (marginally updated paperback version, 2012).

[28] F. Calogero. "Two new solvable dynamical systems of goldfish type." *J. Nonlinear Math. Phys.* **17**: 397–414. (2010). DOI: 10.1142/S1402925110000970.

[29] F. Calogero. "Discrete-time goldfishing." *SIGMA* **7**: 082, 35 pages (2011).

[30] F. Calogero. "A new goldfish model." *Theor. Math. Phys.* **167**: 714–724. (2011).

[31] F. Calogero. "On a technique to identify solvable discrete-time many-body problems." *Theor. Math. Phys.* **172**: 1052–1072. (2012).

[32] F. Calogero. "New solvable many-body model of goldfish type." *J. Nonlinear Math. Phys.* **19**: 1250006. (19 pages) (2012). DOI: 10.1142/S1402925112500064.

[33] F. Calogero. "Another new solvable model of goldfish type." *SIGMA* **8**:046 (17 pages) (2012).

[34] F. Calogero. "New solvable variants of the goldfish many-body problem", *Studies Appl. Math.* **137**(1), 123-139 (2016); DOI: 10.1111/sapm.12096.

[35] F. Calogero. "Nonlinear differential algorithm to compute all the zeros of a generic polynomial." *J. Math. Phys.* **57**: 083508 (4 pages) (2016). DOI: 10.1063/1.4960821.

[36] F. Calogero. "Comment on 'Nonlinear differential algorithm to compute all the zeros of a generic polynomial' [*J. Math. Phys.* **57**: 083508 (2016)]." *J. Math. Phys.* **57**: 104101 (4 pages) (2016).

[37] F. Calogero. "A solvable N-body problem of goldfish type featuring N^2 arbitrary coupling constants." *J. Nonlinear Math. Phys.* **23**: 300–305 (2016). DOI: 10.1080/14029251.2016.1175823.

[38] F. Calogero. "Novel isochronous N-body problems featuring N arbitrary rational coupling constants." *J. Math. Phys.* **57**: 072901 (2016). DOI:10.1063/1.4954851.

[39] F. Calogero. "Three new classes of solvable N-body problems of goldfish type with many arbitrary coupling constants." *Symmetry* **8**: 53. (2016). DOI:10.3390/sym8070053.

[40] F. Calogero. "Yet another class of new solvable N-body problems of goldfish type." *Qualit. Theory Dyn. Syst.* **16**: (3), 561–577 (2017). DOI: 10.1007/s12346-016-0215-y.

[41] F. Calogero. "New solvable dynamical systems." *J. Nonlinear Math. Phys.* **23**: 486–493. (2016).

[42] F. Calogero. "*Integrable Hamiltonian N*-body problems of goldfish type featuring *N arbitrary* functions." *J. Nonlinear Math. Phys.* **24**(1): 1–6. (2017).

[43] F. Calogero. "New C-integrable and S-integrable systems of nonlinear partial differential equation." *J. Nonlinear Math. Phys.* **24**(1): 142–148. (2017).

[44] F. Calogero. "Novel differential algorithm to evaluate *all* the zeros of any *generic* polynomial." *J. Nonlinear Math. Phys.* **24**: 469–472. (2017).

[45] F. Calogero. "Zeros of *entire* functions and related systems of *infinitely many* nonlinearly coupled evolution equations." *Theor. Math. Phys.* **196** (2), 1111–1128 (2018).

[46] F. Calogero and A. Degasperis. *Spectral Transform and Solitons: Tools to Solve and Investigate Nonlinear Evolution Equations*. North Holland: Amsterdam. (1982). pp.514.

[47] F. Calogero and A. Degasperis. "On the quantization of Newton-equivalent Hamiltonians." *Amer. J. Phys.* **72**: 1202–1203. (2004).

[48] F. Calogero and S. De Lillo, "The Eckhaus PDE $iy_t + y_{xx} + 2(|y|^2)_x y + |y|^4 y = 0$." *Inverse Problems* **3**: 633–681. (1987); Erratum **4**: 571. (1988).

[49] F. Calogero and J.P. Françoise. "Hamiltonian character of the motion of the zeros of a polynomial whose coefficients oscillate over time." *J. Phys. A: Math. Gen.* **30**: 211–218. (1997).

[50] F. Calogero and J.P. Françoise. "A novel solvable many-body problem with elliptic interactions." *Int. Math. Res. Notices* **15**: 775–786. (2000).

[51] F. Calogero, J.P. Françoise, and M. Sommacal. "Solvable nonlinear evolution PDEs in multidimensional space involving trigonometric functions." *J. Phys. A: Math. Theor.* **40**: F363–F368. (2007).

[52] F. Calogero, J.P. Françoise, and M. Sommacal. "Solvable nonlinear evolution PDEs in multidimensional space involving elliptic functions." *J. Phys. A: Math. Theor.* **40**: F705–F711. (2007).

[53] F. Calogero and D. Gómez-Ullate. "Two novel classes of solvable many-body problems of goldfish type with constraints." *J. Phys. A: Math. Theor.* **40**: 5335–5353. (2007).

[54] F. Calogero and D. Gómez-Ullate. "Asymptotically isochronous systems." *J. Nonlinear Math. Phys.* **15**: 410–426. (2008).

[55] F. Calogero, D. Gómez-Ullate, P. M. Santini, and M. Sommacal. "The transition from regular to irregular motions, explained as travel on Riemann surfaces." *J. Phys. A: Math. Gen.* **38**: 8873–8896. (2005). arXiv:nlin.SI/0507024 v1 13 Jul 2005.

[56] F. Calogero, D. Gómez-Ullate, P. M. Santini, and M. Sommacal. "Towards a theory of chaos explained as travel on Riemann surfaces." *J. Phys. A.: Math. Theor.* **42**: 015205 (26 pages) (2009). DOI: 10.1088/1751-8113/42/1/015205.

[57] F. Calogero and S. Graffi. "On the quantization of a nonlinear Hamiltonian oscillator." *Phys. Lett. A* **313**: 356–362. (2003).

[58] F. Calogero and S. Iona. "Novel solvable extensions of the goldfish many-body model." *J. Math. Phys.* **46**: 103515 (2005).

[59] F. Calogero and E. Langmann. "Goldfishing by gauge theory." *J. Math. Phys.* **47**, 082702: 1–23 (2006).

[60] F. Calogero and F. Leyvraz. "On a class of Hamiltonians with (classical) isochronous motions and (quantal) equispaced spectra." *J. Phys. A: Math. Gen.* **39**: 11803–11824. (2006).

[61] F. Calogero and F. Leyvraz. "Isochronous extension of the Hamiltonian describing free motion in the Poincaré half-plane: classical and quantal treatments." *J. Math. Phys.* **48**, 092903: 1–15. (2007).

[62] F. Calogero and F. Leyvraz. "On a new technique to manufacture isochronous Hamiltonian systems: classical and quantal treatments." *J. Nonlinear Math. Phys.* **14**: 505–529. (2007).

[63] F. Calogero and F. Leyvraz. "General technique to produce isochronous Hamiltonians." *J. Phys. A: Math. Theor.* **40**: 12931–12944. (2007).

[64] F. Calogero and F. Leyvraz. "New solvable discrete-time many-body problem featuring several arbitrary parameters." *J. Math. Phys.* **53**: 082702 (19 pages) (2012).

[65] F. Calogero and F. Leyvraz. "New solvable discrete-time many-body problem featuring several arbitrary parameters. II." *J. Math. Phys.* **54**: 102702 (15 pages) (2013). DOI: http://dx.doi.org/10.1063/1.4822419.

[66] F. Calogero and F. Leyvraz. "A nonautonomous yet solvable discrete-time *N*-body problem." *J. Phys. A: Math. Theor.* **47**, 105203 (10 pages) (2014).

[67] F. Calogero and F. Leyvraz. "The peculiar (monic) polynomials the zeros of which equal their coefficients." *J. Nonlinear Math. Phys.* **24**: 541–551 (2017). DOI:10.1080/14029251.2017.1375690.

[68] F. Calogero and F. Leyvraz. "Examples of Hamiltonians isochronous in configuration space only, and their quantization." *J. Math. Phys.* **59**(6), 062701 (7 pages) (2018), DOI: 10.1063/1.5010590.

[69] F. Calogero and M. Sommacal. "Solvable nonlinear evolution PDEs in multidimensional space." *SIGMA* **2**: 088 (2006) (17 pages), nlin.SI/0612019.

[70] F. Calogero and Ge Yi. "A new class of solvable many-body problems." *SIGMA* **8**: 066 (29 pages) (2012).

[71] A. J. Di Scala and O. Maciá. "Finiteness of Ulam polynomials." La matematica e le sue applicazioni, Quaderni del Dipartimento di Matematica, Politecnico Torino, **11**: July 2008. Turin, Italy; arXiv:00904.0133v1[math.AG]1April2009.

[72] A. Erdélyi (editor). *Higher transcendental functions*, vol. II. McGraw-Hill: New York. (1953).

[73] G. Gallavotti and C. Marchioro. "On the computation of an integral." *J. Math. Anal. Appl.* **44**: 661–675. (1973).

[74] C. S. Gardner, J. M. Greene, M. D. Kruskal, and R. M. Miura. "Method for Solving the Korteweg-deVries Equation." *Phys. Rev. Lett.* **19**: 1095–1097 (1967). DOI: 10.1103/PhysRevLett.19.1095.

[75] D. Gómez-Ullate, P. M. Santini, M. Sommacal, and F. Calogero. "Understanding complex dynamics by means of an associated Riemann surface." *Physica D.* **241**: 1291–1305: (2012).

[76] D. Gómez-Ullate and M. Sommacal. "Periods of the goldfish many-body problem." *J. Nonlinear Math. Phys.* **12**, Suppl. **1**: 351–362. (2005).

[77] P. G. Grinevich and P. M. Santini. "Newtonian dynamics in the plane corresponding to straight and cyclic motions on the hyperelliptic curve $\mu^2 = \nu^n - 1$, $n \in \mathbb{Z}$: Ergodicity, isochrony and fractals." *Physica D* **232**: 22–32. (2007).

[78] A. Guillot. "The Painleve' property for quasi homogeneous systems and a many-body problem in the plane." *Comm. Math. Phys.* **256**: 181–194. (2005).

[79] U. Jairuk, S. Yoo-Kong, and M. Tanasittikosol. "On the Lagrangian structure of Calogero's goldfish model." arXiv:1409.7168 [nlin.SI] (2014).

[80] A. Kundu. "Landau–Lifshitz and higher-order nonlinear systems gauge generated from nonlinear Schrödinger-type equations." *J. Math. Phys.* **25**: 3433–3438. (1984).

[81] F. Leyvraz. "An approach for obtaining integrable Hamiltonians from Poisson-commuting polynomial families." *J. Math. Phys.* **58**: 072902 (2017).

[82] *Mathematica*, Wolfram Research, Inc., Version 11.0. Champaign, IL, USA. (2016).

[83] V. G. Marikhin. "Electron-levels dynamics in the presence of impurities and the Ruijsenaars-Schneider model." *JETP Lett.* **77**:44 (2003). DOI:10.1134/1.1561980.126

[84] J. Moser. "Three integrable Hamiltonian systems connected with isospectral deformations." *Adv. Math.* **16**: 197–220. (1975).

[85] M. C. Nucci. "Calogero's 'goldfish' is indeed a school of free particles." *J. Phys. A: Math. Gen.* **37**: 11391–11400. (2004).

[86] M. A. Olshanetsky and A. M. Perelomov. "Explicit solution of the Calogero model in the classical case and geodetic flows on symmetric spaces of zero curvature." *Lett. Nuovo Cimento* **16**: 333–339. (1976).

[87] G. Salmon. *Lessons Introductory to the Higher Modern Algebra.* Cambridge University Press: Cambridge, U. (1885).

[88] P. C. Sabatier (ed.). *Inverse Methods in Action.* Springer: Berlin. (1990).

[89] M. Sommacal. "The transition from regular to irregular motion, explained as travel on Riemann surfaces." PhD thesis, SISSA, Trieste. (2005).

[90] P. R. Stein. "On polynomial equations with coefficients equal to their roots." *A. Math. Monthly* **73**(3): 272–274. (1966).

[91] Y. B. Suris. "Time discretization of F. Calogero's 'Goldfish'. *J. Nonlinear Math. Phys.* **12**, Suppl. **1**: 633–647. (2005).

[92] S. M. Ulam. *A collection of mathematical problems.* Interscience: New York. 1960 (see pages 30–31).

[93] V. E. Zakharov. "On the dressing method." in [88], pp. 602–623 (see p. 622).